Weather Rage

Science Spectra Book Series

Series Editor: Vivian Moses, King's College, University of London, UK

Volume 1
The Key to the Future
By John Cater

Volume 2
On Beyond Uranium
By Sigurd Hofmann

Volume 3
A Terrible Beauty is Born
By Brendan Curran

Volume 4
The Subtle Beast
By André Ménez

Volume 5
Weather Rage
By Ross Reynolds

Weather Rage

Ross Reynolds

LONDON AND NEW YORK

First published 2003
by Taylor & Francis
11 New Fetter Lane, London EC4P 4EE

Simultaneously published in the USA and Canada
by Taylor & Francis Inc,
29 West 35th Street, New York, NY 10001

Taylor & Francis is an imprint of the Taylor & Francis Group

Typeset in Optima by Wearset Ltd, Boldon, Tyne and Wear
Printed and bound in Malta by Gutenberg Press Ltd

British Library Cataloguing in Publication Data
A catalogue record for this book is available from the British Library

Library of Congress Cataloging in Publication Data
Weather rage / edited by Ross Reynolds
p. cm. – (Science spectra)
Includes bibliographical references and index.
1. Severe storms. I. Reynolds, Ross. II. Series.
QC941 .W43 2003
551.55–dc21

2002013372

ISBN 0-415-36981-9 (hbk)
ISBN 0-415-36982-7 (pbk)

Cover image (lightning) courtesy of NOAA Photolibrary, NOAA Central Library; OAR/ERL/National Severe Storms Laboratory (NSSL).

Contents

1

Wind, cloud, rain and blizzard

An introduction

The notion of 'raging' weather conjures up an immediate image of strong wind as an important ingredient, possibly accompanied by rain, snow or hail. Thunder and lightning also 'rage' but do not have to be accompanied by wind to qualify. A thesaurus offers synonyms like 'frenzy', 'fury', 'rampage', 'storm' and 'surge' that lend themselves as pointers to the kind of weather falling into the category of raging; they help to convey a sense of what it can be like.

Some atmospheric disturbances produce the kind of extreme weather events making the national and international news headlines on a regular basis. Just about everyone is interested to know more about hurricanes, typhoons and tornadoes. There is very often confusion about whether or not they are the same thing and whether, for example, we can control these extremely violent phenomena. In the higher latitudes, wild weather is most often linked to the passage of travelling 'frontal' depressions frequently bringing widespread strong winds and heavy rain.

Thunder and lightning affect many more people but how many of us understand what generates such massive electrical discharges and why they occur more at some times of the year than at others?

This book will explain the origin and nature of different types of wild weather observed across many areas of the world – and those that occur on different 'scales'. They range from the tornado, locally devastating over short periods, to typhoons or hurricanes that can wreak havoc over extensive areas for several days. In this chapter we will survey some of the main factors underlying weather and then go on to look at some of their consequences in more detail.

Why does the wind blow?

The first thing to realise is that wind is quite simply the air in motion. That seems obvious; what is not so obvious is that there is an enormous amount of mass on the move when the wind blows. At or near sea-level, each cubic

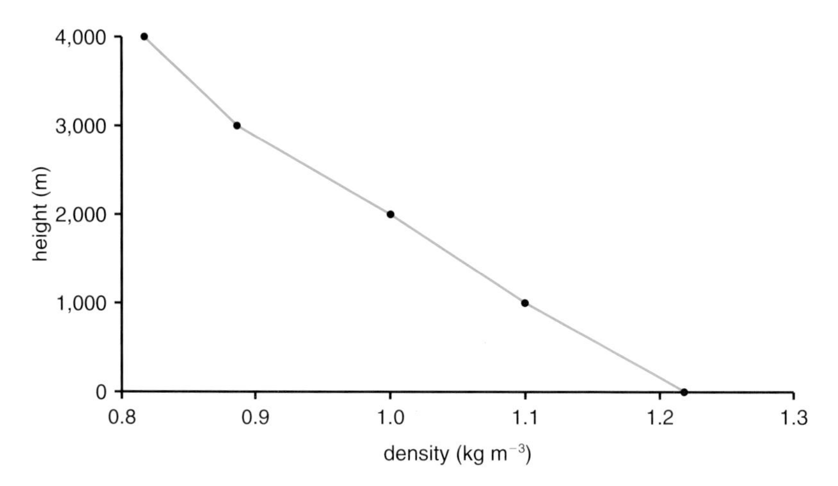

Figure 1.1 Air density versus height above sea level.

metre of air weighs about 1.2 kilograms (equivalent to 1.2 ounces per cubic foot). The air's density decreases quickly upwards away from the surface (Figure 1.1) as the atmospheric pressure falls off very rapidly in that direction.

To understand why the air moves in the first place, we need to start with the concept of atmospheric pressure measured routinely by barometers. We are used to seeing weather maps on television or in newspapers with their familiar 'isobars', lines that connect points where the mean sea-level atmospheric pressure is the same. *Iso* means equal and *bar* comes from the Greek word *baros* which means weight.

So, when we measure pressure, we are also measuring the weight of the atmosphere above the barometer, whether it be an official mercury one as used in weather services or the circular 'aneroid' design more common in our homes. They tell us how many inches or millimetres of mercury the pressure is equivalent to – or how many millibars (mbar) – a 'bar' is an atmospheric pressure of 760 millimetres of mercury; a millibar is 1/1000th of that value. We are literally weighing the ocean of air that sits above the barometer at a particular instant in time, an ocean on the bottom of which we live (Figure 1.2).

Highs and lows on the maps

High pressure on a weather map means that there is a greater mass of air above that region than there is above an area of low pressure. There is, in fact, no magical value which separates high from low pressure; it is always simply the relative values which determine where the high and low centres are sited. On a

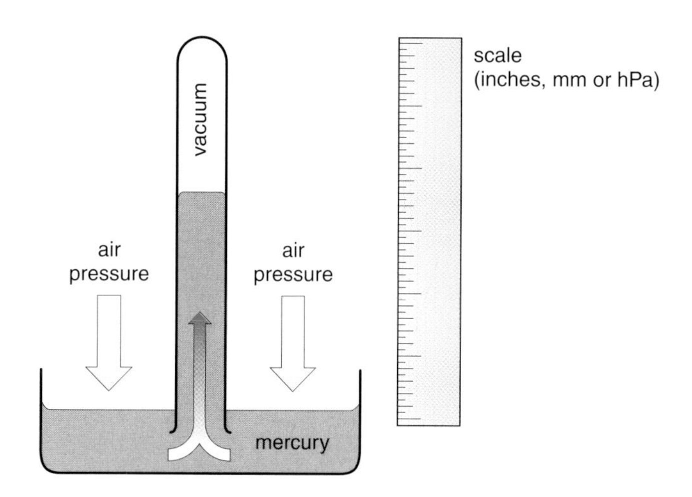

Figure 1.2 The principle of the mercury barometer.

particular day, the highest high might have a centre of 1015 mbar and the lowest low one of 1002 mbar. A different day in the same region might see the centres with a high and low of 1045 and 1017 mbar respectively.

Indeed, it is possible to calculate the mass of air above a barometer from the pressure it records. You can do so by dividing the barometric pressure by the value of gravity. If we assume a standard value for gravity at mean sea-level, then a pressure of, say, 1040 mbar equates to 10.60 tonnes of air above every square metre of the Earth's surface, while one of 960 mbar converts to 10.20 tonnes. This means quite literally that typically, with the global average atmospheric pressure at sea-level of 1013.2 mbar, we have 10.33 tonnes of air above every square metre at the sea surface. This works out at around a quarter of a tonne above an adult head (Figure 1.3). That is a lot of air; no wonder its effects become noticeable when it is on the move. You don't need to worry about being crushed by such a weight of air; it is exactly balanced by the pressure within our bodies so we are in no danger either of being squashed or of bursting.

Pressure is useful

What does pressure mean for the weather? When we look at a weather map with isobars on it, regions of high pressure are those above which there is a high 'head' of air (in the same way as we can imagine a high head of water in other situations). In contrast, regions of low pressure are those underneath a relatively low head of air (Figure 1.4).

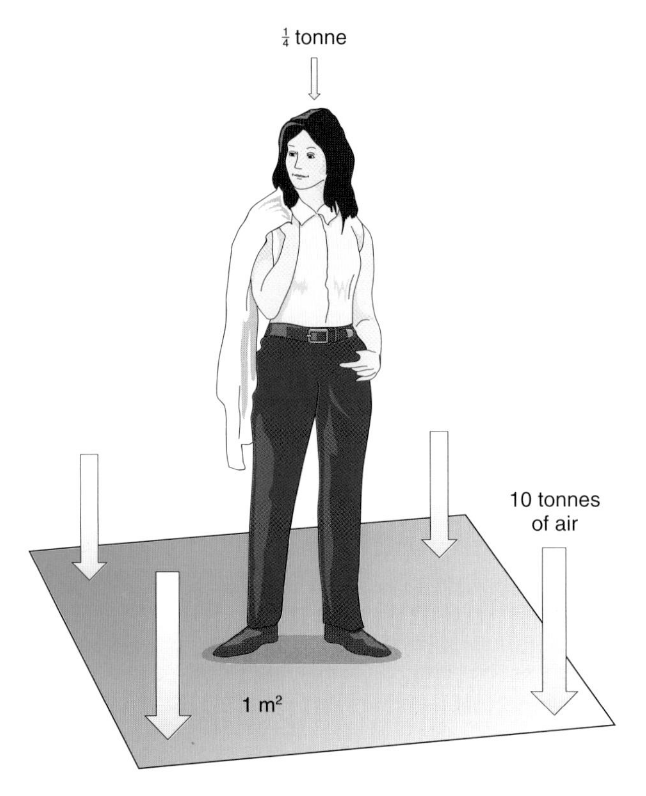

Figure 1.3 The approximate weight of air above 1 square metre of ground and on a human head, both at sea level.

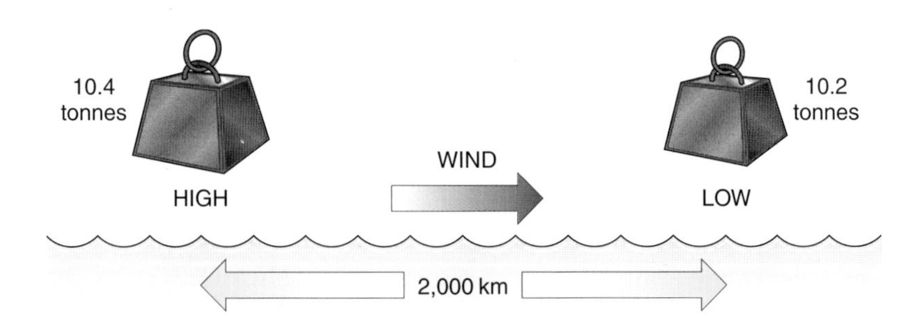

Figure 1.4 Differences in the masses of air in regions of high and low pressure.

Air is a fluid and, as such, is forced to flow from regions of high pressure towards those of low pressure. How fast the air will flow depends on the size of the difference in the pressure centres and the distance separating them. A large difference in pressure over a short distance means a steep gradient of pressure on the weather map, while a small difference over a larger distance implies a much weaker gradient (Figure 1.5). It is rather like water flowing down a slope – the steeper the slope, the faster the water flows but over the Earth's surface, of course, the air flows horizontally, not down a slope. Areas on a weather map with a rapid change of pressure across the surface (i.e. with closely packed isobars) will experience air flowing rapidly (strong winds) while places with weak pressure gradients (isobars spread well apart) will see lighter winds.

Wind strength

Most of us have experienced wind speeds varying from nil or calm to perhaps gale force. The strongest winds in nature are those associated with the incredibly steep horizontal pressure gradients across a tornado. Hurricanes and typhoons also have gradients that are steep but not so strong as the ones in tornadoes. Frontal depressions are the features that produce gale, storm and occasionally hurricane force winds in middle latitudes.

The terms 'gale force' or 'storm force' are, of course, really quite old. They date back some two hundred years to the time when Admiral Beaufort of the Royal Navy needed to develop a scale useful to men-of-war. It was important to set the sails effectively to the given weather, so his scale related wind strength to the state of the sea surface. The original descriptions are still used for sailing;

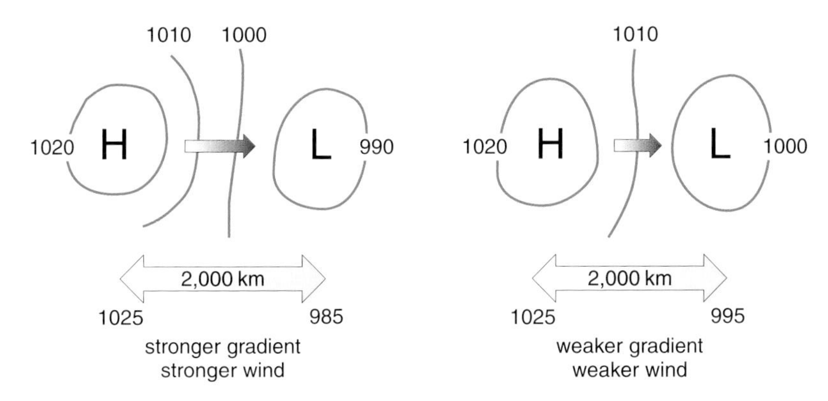

Figure 1.5 The link between the horizontal pressure gradient and the resultant wind speed.

those summarising what happens on land in different wind forces were added later. The real driving force of the air that creates the wind is the horizontal change in pressure: the steeper it is, the stronger the wind. This means that the Beaufort number increases with wind strength; the numbers have stood the test of two centuries and are familiar to most of us even today (Table 1.1).

Some confusion – particularly in the press – can ensue when it might be reported that there have been hurricane force gusts in, say, western Scotland or the south coast of the Isle of Wight. That term relates only to the speed of the gust and does not mean that a hurricane is crossing the British Isles! We never see hurricanes as full-blown systems in western Europe but in the late summer or early autumn we do on occasion see their remnants.

Table 1.1 Beaufort scale for the standard height of 10 metres (33 feet) at a well-exposed site

Beaufort force	**description**	**Wind speed**	
		metres per second	**knots**
0	calm, smoke rises vertically	<0.6	<1
1	light air, wind direction shown by smoke but not by vane	0.6–1.7	1–3
2	slight breeze, wind felt on face, leaves rustle, vane moves	1.8–3.1	4–6
3	gentle breeze, leaves and small twigs move constantly	3.2–5.4	7–10
4	moderate breeze, raises dust and loose paper, small branches move	5.5–8.1	11–16
5	fresh breeze, small trees in leaf sway, crested wavelets on inland waters	8.2–10.8	17–21
6	strong breeze, large branches move, whistling in telephone wires	10.9–13.9	22–27
7	near gale, whole trees move, affects walkers noticeably	14.0–16.8	28–33
8	gale, twigs break off trees, impedes progress in general	16.9–20.6	34–40
9	strong gale, slight structural damage e.g. roof tiles/slates	20.7–24.2	41–47
10	storm, trees uprooted, considerable damage to structures	24.3–28.3	48–55
11	violent storm, very rare, damage widespread	28.4–32.6	56–64
12	hurricane, violence and destruction	>32.6	>64

However, they are very much less windy and less wet by the time they reach Europe. Matters are quite different in the Caribbean or the south-eastern part of the United States for example; there hurricanes are experienced most years, some causing deaths and enormous amounts of damage.

If the wind reaches force 12 on the Beaufort scale it is officially rated as 'hurricane force' even though not necessarily produced by a hurricane which is a particular weather system we will discuss in more detail in Chapter 2. It is, however, the criterion that is used to this day to decide whether a cyclone (a low pressure system) in the tropics has become a hurricane (or typhoon or cyclone; the name used depends on the ocean basin). If the surface wind, averaged over a few minutes, exceeds 64 knots then it has made the grade.

Mass on the move

The flow of air is enormous. We can imagine a situation with a westerly wind with a speed of 10 metres a second (about 20 knots), blowing in a layer 1 km (3,300 feet) deep stretching from the north of Scotland to the south of England, a distance of about 900 km or 560 miles (Figure 1.6).

We can work out how much air flows across this north–south 'wall' 1 km high by multiplying the wind speed (10 metres a second) by the area it is blowing across (1 km × 900 km) and by the density of the air (say 1 kilogram per cubic metre in the lowest 1 km of the atmosphere). This tells us that, every second, some 9 million tonnes of air flows through the 'wall'!

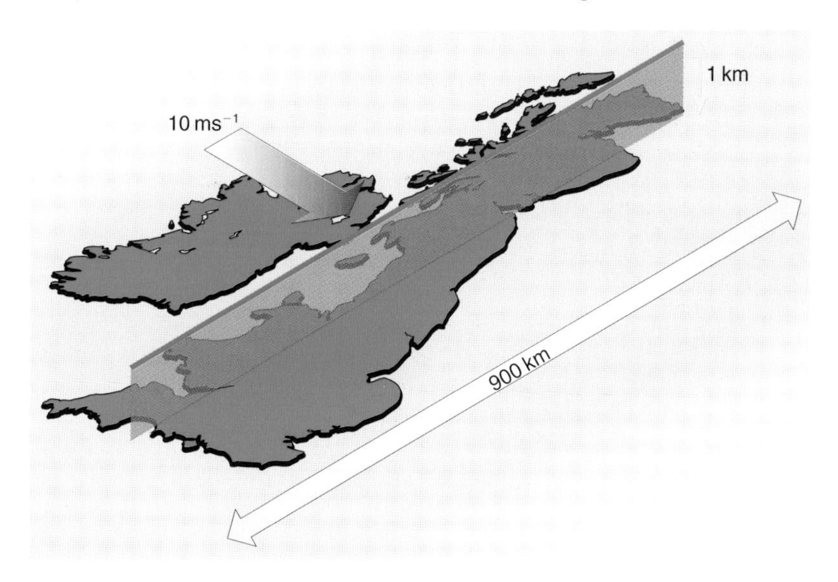

Figure 1.6 A wall of air along the British Isles.

With all this mass of air moving around, it is not surprising that the pressure in the highs and lows changes all the time. The air has to flow out of highs down the pressure gradient into lows, which means that the surface high pressure should decrease (because mass is being removed by the wind) while in the lows it should increase as they fill up with air coming from the highs. The problem in nature is that, although mass is being transferred in this way at the surface, it is also being transferred throughout the depth of the atmosphere. To understand surface pressure changes we therefore have to know what is going on with the wind patterns in depth.

Although the wind blows because it is transporting mass from high to low pressure areas, it does not, however, blow directly from a high to a low at right angles to the isobars; because we move round on a rotating Earth, we see the air spiral out of highs and spiral into lows at a narrow angle to the isobars (Figure 1.7).

Damaging force

The fact that large amounts of air are on the move means that extreme winds apply forces able to cause very significant damage to natural and artificial structures. We can see why that happens if we convert the strength of the wind at sea-level into a mass pressing a vertical surface as the air blows at it. In a light wind this is very small but, for a whole gale (Force 10), it can rise to more than 50 kilograms per square metre. This means that a mass of about one tonne of air is imposed on a wall 2 metres high and 10 metres long (6.5 feet × 32.8 feet).

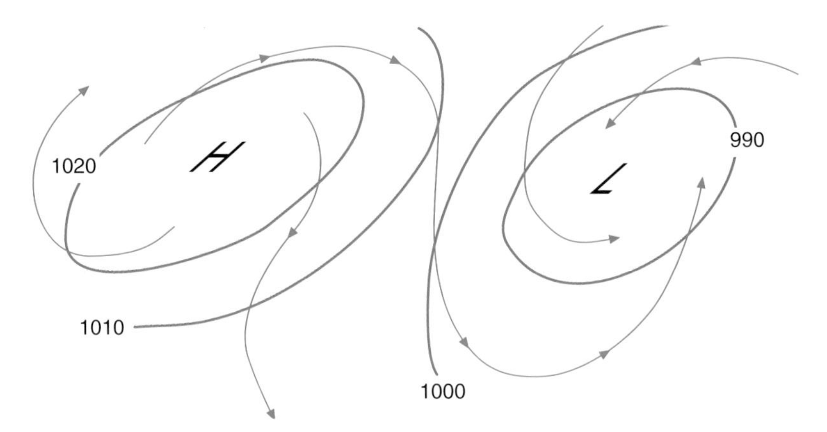

Figure 1.7 Winds spiralling around highs and lows: surface/low-level air spiralling into lows and out of highs.

It is often the short-lived gusts that produce the structural damage associated with extremely windy conditions. How gusty the flow is depends in fact on a number of factors, including how 'rough' the surface of the Earth is. The rougher it is, the more gusty the wind is likely to be; in built-up areas, for example, the surface is rougher than out in the countryside. Moreover, it is more gusty on a day when there are deep up and down currents in the atmosphere; the way in which the wind speed changes with height in the lowest few kilometres of the atmosphere also affects its gustiness.

Deep ascent and descent

Air at and near the surface tends to rise as it flows into low pressure areas while air flowing out of highs has descended. This is of great consequence for our day-to-day weather because when air rises it cools by expanding and when it sinks it is warmed by being compressed; these are called *adiabatic temperature changes* (Figure 1.8).

The process of cooling is critical for the creation of clouds and therefore precipitation when it happens to air that is damp. If the air swirling into a low is dry, it may not produce cloud at all. By contrast, air subsiding within high pressure areas tends to suppress deep cloud and precipitation. That is why highs are often associated with generally dry weather, although conditions can be miserable if they have extensive layer cloud within their circulation. Highs do not by any means always mean unbroken sunshine and clear nights!

Creating cloud

The ascent of air within low pressure systems leads to the creation of widespread cloud which normally occurs as deep, extensive layers. Quite often the lows are 'frontal' which means they have attendant warm, cold and occluded fronts – of which more later.

Rising air not saturated with water vapour (i.e. effectively cloud-free air) experiences continuously falling pressure as it ascends. Because this air is moving into an area of lower pressure it expands which, in turn, means that

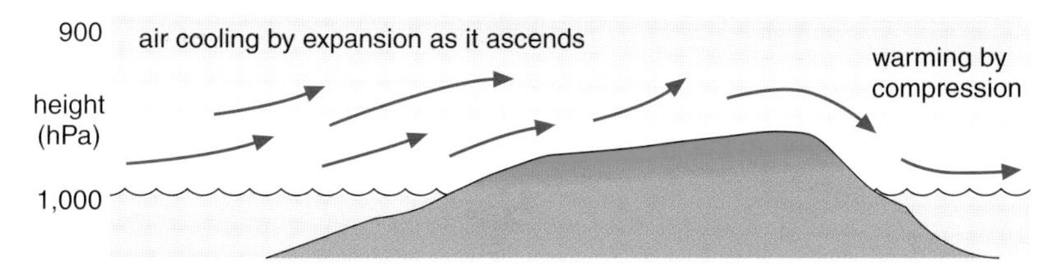

Figure 1.8 Adiabatic cooling of ascending, and warming of descending air.

it cools. In fact, the rate of cooling is quite rapid: it is 1°C for every 100 metres of ascent so if the air caught up in this large-scale motion is quite damp, with a high relative humidity, the cooling will quickly lead to widespread condensation and an extensive layer of 'stratiform' cloud. This type of layer cloud is formed by gentle ascent over very large areas; it can easily stretch many hundreds of kilometres across the Earth's surface.

The droplets making up the cloud are extremely small, with an average diameter of about 50 micrometres (50 millionths of a metre, equal to less than a millionth of an inch). In contrast to this, small and large raindrops have diameters of around 1,000 and 5,000 micrometres (1 and 5 mm – about 1/25 to 1/5 of an inch), respectively. Drizzle drops are about 200 micrometres in diameter. This means that a typical cloud droplet needs somehow to increase its volume about 60-fold in order to become even an average drizzle drop, let alone become a raindrop.

Settling drops

The terminal speed of a 'typical' cloud droplet settling under gravity is around 8 centimetres (3 inches) a second: however, the rate of fall under the force of gravity is balanced by frictional drag imposed on the drop by the air through which it is falling. If these drops are present in air ascending faster than 8 centimetres a second – as is the case in widespread ascent – the drops rise within the cloud. In fact, the very existence of clouds tells us that the air is ascending inside them at different rates, differences which are reflected in the horizontal dimension of the cloud. Thus, sheets of cloud stretching across many hundreds or a thousand or more kilometres are borne on relatively gently rising air going up at a few tens of centimetres a second. By contrast, smaller clouds such as cumulus, which are growing strongly upwards, are carried on more powerful currents, typically a few to some tens of metres per second depending on their vigour.

A large raindrop has a terminal fall speed of about 9 metres (28 feet) a second, reaching a limit, when around 7,000 micrometres (7 mm or about $\frac{1}{4}$ inch) in diameter, of some 10 metres (33 feet) a second. At that speed it flattens out so significantly that it breaks up into a number of small drops; thus there is an upper limit to the sizes of the raindrops we actually see in nature.

Creating rain and snow

It might surprise you to know that cloud droplets do not freeze at 0°C. In fact, even at a temperature of −10°C, only one in a million of them will actually be frozen. At −30°C it is still only about one in a thousand and not until the temperature falls to −40°C are they all ice crystals! The liquid water droplets in such clouds are termed 'supercooled'.

The fact that clouds are frequently composed of both liquid water droplets and ice crystals is extremely significant for the way in which precipitation is formed. If ice crystals and liquid droplets occur together in the same volume of cloud, the crystals grow at the expense of the droplets. This is because what is called the 'saturation vapour pressure' is lower over ice than over liquid water surfaces; the upshot is that any water vapour in the vicinity will condense onto the ice crystals rather than on the water droplets (just as in a freezer: once ice forms inside, for whatever reasons, it tends to accumulate more). The droplets can actually become smaller as they lose water by evaporation. The result is that in such cloud, the ice crystals grow at the expense of the droplets through the so-called *Bergeron-Findeisen process* named after the two scientists whose work explained the mechanism. It is in this way that most precipitation is formed outside the tropics.

Once the ice crystals do begin to grow, they start to settle out through the cloud and keep growing by coalescence with smaller drops in their path which are essentially 'swept up'. If the temperature within and below the cloud all the way down to the Earth's surface is less than 0°C, the fully-formed aggregated ice crystals fall out as snow. If the temperature is above freezing just in the lowest layers, we might at the ground see sleet which is simply melting snow. However, when the temperature is above freezing through a substantial part of the lower reaches of the cloud, rain will fall to the surface. All the rain we see outside the tropics started life as ice crystals and snowflakes in the higher, colder reaches of the cloud.

What makes torrential rain?

Really heavy rain needs particular ingredients. The greatest falls over short periods are always linked to deep convective clouds – the sort of cumulus clouds that have the brilliant 'cauliflower' tops or the even more spectacular icy anvils flowing out of the top of cumulonimbus. Sometimes the cumulonimbus is embedded in an extensive layer cloud so that we cannot see their brilliant tops; people on the ground may experience them only by a significant increase in rain rate.

Heavy rain involves very moisture-rich air feeding into the base of the cloud. One can estimate how moisture-rich it is by noting the air's dewpoint or wetbulb temperature (Figure 1.9), both measures of the absolute amount of water vapour present (i.e. the number of grams of water vapour present in a kilogram of air). Because evaporation is greater in the summer than winter, the potential for large amounts of rain is higher in the warmer season. Of course, it is always high in the humid tropics where the specific humidity is substantial year round.

When this very humid air ascends into deep cumulus clouds, it leads to

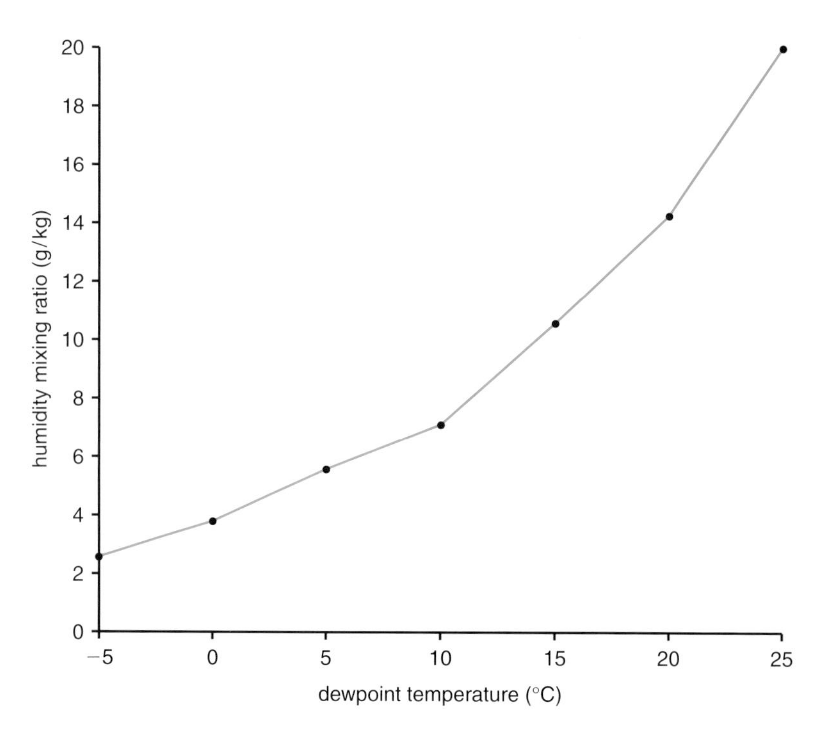

Figure 1.9 Dewpoint versus water vapour concentration.

very high liquid water content within them. Furthermore, such warm season clouds tend to contain more vigorous updrafts because the surface heating that sparks them off is itself more substantial. This means that precipitation-size drops which may begin to fall through the cloud can, in practice, be swept back up again by the powerful ascending currents. Larger drops grow by coalescence both when falling and also when ascending.

Within the convective cloud, growth like this with strong updrafts carrying water-rich air is the kind of environment that produces intensely heavy rain at the surface. All the record short-term falls recorded are from such clouds and tend to occur in the warm season.

The top three twenty-four hour totals ever recorded in the British Isles were all in the summer:

1.	279.4 mm	Martinstown, Dorset	18 July 1955
2.	242.8 mm	Bruton, Somerset	28 June 1917
3.	241.3 mm	Upwey, Dorset	18 July 1955

Of the top sixteen such falls, twelve happened in June, July or August.

The twenty-four hour record for the United States was reported on 25 July 1979 at Alvin, Texas where 1092.2 millimetres (43 inches) fell! The one-minute record comes from Unionville, Maryland where a torrent produced a total of 30.5 millimetres (1.2 inches). Similarly, the top three short-period falls in the UK were: 31.7 mm in 5 minutes at Preston in Lancashire (10 August 1893), 50.0 mm in 10 minutes at Wisbech, Cambridgeshire (28 June 1970) and 55.9 mm in 15 minutes at Bolton, Greater Manchester (18 July 1974). Twenty of the top twenty-four short duration falls took place in June, July or August.

The record twenty-four hour falls were associated with slow-moving deep convective clouds forming and dying out just about continuously round the clock. The reason that the really intense falls of short duration are so short lived is because it is not possible for such a rate to be maintained for many hours – the water cannot be cycled rapidly enough through the cloud to keep such extremely intense falls going for very long.

What makes blizzards?

Blizzard conditions are those of *blowing* snow which does not have actually to be falling at the time. The crucial ingredient is strong wind which either whips up snow from the surface or combines with falling snow to produce the blizzard.

Snow showers, like rain or hail showers, fall as short-lived bouts of precipitation from travelling cumulus clouds. Any blizzard-like conditions associated with them tend, of course, to be short-lived. The full-blown blizzards are longer lived and linked to larger-scale disturbances like travelling low pressure systems. The recipe is for such lows to produce widespread snowfall and to be 'deep' (i.e. low pressure) so that their large horizontal pressure gradients provide the all-important winds to blow the snow around strongly as it falls. But, as we mentioned earlier, blizzards can arise simply from lying snow being whipped up from the surface.

Paradoxically perhaps, one important component of snowfalls is relatively mild air. This does not mean that the temperature has to be above 0°C but neither should it be frigidly cold. This is because extremely cold air has a very small capacity for holding water vapour – so small, in fact, that even if cloud could be squeezed out of it, hardly any snow would ensue. To produce a really substantial snowfall, relatively warm, moisture-rich air needs to be caught up in the travelling lows that are the source of widespread, prolonged precipitation. This is most often the case with middle-latitude frontal depressions within the circulation of which are extensive, deep layers of air originating over the subtropical oceans. It is this air moving towards the pole and upwards over the fronts that produces widespread cloud and snow. The massive falls over the Alps during the later part of the winter of 1998/1999 came from moisture which had evaporated from far-flung warm sea or ocean surfaces, probably many hours or even days earlier.

Atmospheric layering

It is important to know some basic terminology that will be used quite frequently in the discussions which follow in subsequent chapters.

The atmosphere is divided into a number of layers defined on the basis of the way in which the average temperature changes in the vertical direction. The lowest layer, at the bottom of which we live, is the 'troposphere' (Figure 1.10). This word comes from the Greek *tropos* which means turning – signifying that the air within the troposphere is characterised by deep churning motions. This is not always the case – in fact the layer can be quite 'stable' which means that the flow is mainly more-or-less horizontal. However, one important feature of this layer is that it contains virtually all the clouds, pollution and weather.

The troposphere is the layer within which the temperature decreases with height until the 'tropopause' is reached. This is a kind of lid separating the troposphere from the layer above – the stratosphere (Figure 1.10), where

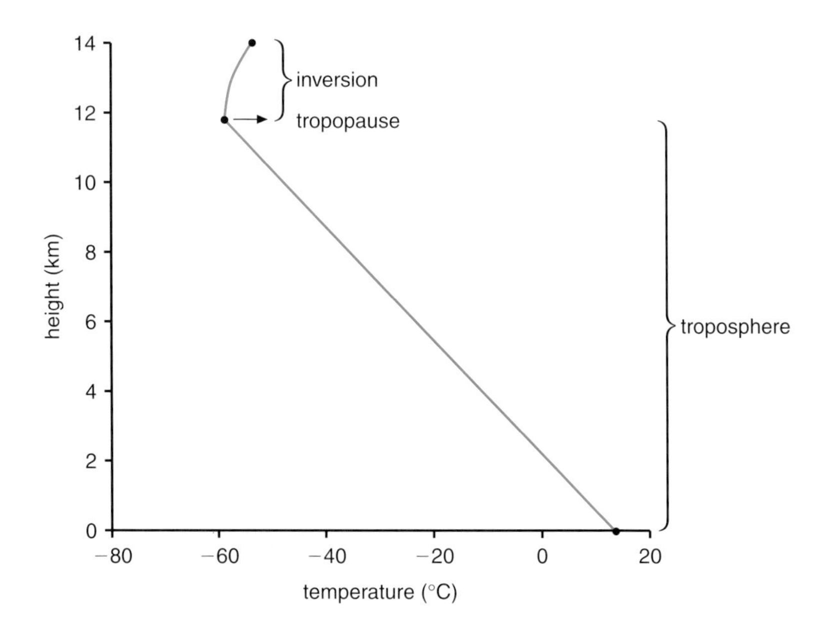

Figure 1.10 Mean vertical temperature profile in the troposphere.

the temperature stops falling and stays the same with increasing height (an *isothermal layer*) or actually increases with height (an *inversion layer*).

The important point about inversion layers (which can and do also exist within the troposphere) is that they are stable because they are marked by temperature increasing with height in the troposphere (i.e. cooler below/warmer above) through a depth of some tens to a few hundred metres. Their presence, provides a 'dampening lid' to any convective motions occurring in the layer below them.

2

Furious tropical winds

Hurricanes, typhoons and cyclones

These three terms all describe the same destructive phenomenon generated over tropical oceans; they not infrequently track into middle latitude regions later in their lives. They are names used regionally: *hurricane* is the title of the feature in the North Atlantic and north-east Pacific Oceans, *typhoon* in the north-west Pacific Ocean and *cyclone* in the South Pacific and Indian Oceans. 'Hurricane' and 'typhoon' are believed to have their possible semantic root respectively in Carib and Chinese, words that mean 'big wind' – not surprisingly! One possible alternative is that 'hurricane' is related to *Huracan*, the God of Evil for one central American tribe.

'Cyclone' is a somewhat ambiguous word because it can mean not only the specific, intense low pressure disturbance described here but also a much more general term used in meteorology to mean a weather-map scale low pressure region. In this more general usage it need be neither tropical nor extremely intense. The word apparently originates from the Greek word *kukloma* meaning wheel.

Definition

The strict definition for hurricane, typhoon, etc. is that it is a large-scale (some hundreds of kilometres across) low pressure system forming over the tropical oceans. It has a core which is warmer than the surrounding atmosphere. Additionally, it has an 'eye' in the centre and spiral rain bands running in towards this centre. The really important criterion is that the *maximum sustained surface winds* (see below for an explanation) must equal or exceed force 12 on the Beaufort scale (i.e. hurricane force) which equates to some 33 metres per second (74 miles per hour).

As stated, hurricanes, typhoons and cyclones are regionally specific names for strong tropical cyclones; a tropical cyclone is the generic term for these non-frontal low pressure systems. Occasionally confusion arises in the sense that vigorous middle-latitude lows in the winter can quite often produce 'hurricane-force' gusts at the surface or even, in rare circumstances, hurricane-force mean winds. This simply means that they have reached force 12 on the Beaufort scale – and not, for example, that we have a hurricane visiting Britain or some other unlikely place!

In their early phase, these tropical systems are classified as tropical depressions, with maximum sustained surface wind speeds lower than 17 metres per second (38 mph). Above that, but weaker than 33 metres per second, they become tropical storms. It is at this stage that they are given a name which stays with them if they intensify into a full-blown hurricane, typhoon, etc.

The definition of 'maximum sustained surface wind speeds' varies somewhat among different authorities. Guidelines issued by the World Meteorological Organisation (a UN Agency) suggest a five-minute average – a standard adopted by the majority of its member states. In contrast, the US National Hurricane Center in Miami and Joint Typhoon Warning Center in Guam both use one-minute averages.

American scale

In the United States the 'Saffir-Simpson' scale is used to convey to meteorologists and the public alike the vigour of a particular hurricane. It was developed in the early 1970s by Herbert Saffir, a consulting engineer in southern Florida, and Robert Simpson, then Director of the National Hurricane Center. Their engineering and meteorological viewpoints were combined to develop the scale which embodies both the type of damage to be expected with different wind speeds and the associated meteorological detail, including the height of the associated storm surge. Wind speed is the critical factor in determining the category because the storm surge is dependent upon the slope of the continental shelf in the region of landfall.

Table 2.1 lists the boundaries for the five categories with a summary of some of the structural and other damage.

Australian scale

The Australian Bureau of Meteorology also uses a scale of 1–5 to convey the severity of tropical cyclones affecting its region (Table 2.2). It is based on the strongest gust and also outlines the type of damage that might be expected within each category.

The where and when

The incidence of tropical cyclones, both seasonally and geographically, gives clues to the factors that are important in their formation. The advent of routine, quality weather satellite images has meant that we have reliable knowledge of the life cycles of all such systems on a global basis, now stretching over three decades.

Table 2.1 Saffir-Simpson Hurricane Intensity Scale

Category	Pressure (hPa)	Mean wind speed (metres/second)	(mph)	Storm surge (metres)
1 (minimal)	more than 980	33–42	74–95	1.2–1.5
No real damage to building structures but mainly to caravans (trailers), shrubs and trees; some coastal road flooding and minor pier damage.				
2 (moderate)	965–979	43–49	96–110	1.6–2.4
Some damage to roofing material, doors and windows; considerable damage to caravans, poorly constructed signs and piers; coastal and low-lying escape routes flood 2–4 hours before arrival of hurricane centre; damage to small craft in unprotected anchorages break moorings.				
3 (extensive)	945–964	50–58	111–130	2.5–4.0
Some structural damage to small residences and utility buildings; damage to shrubs and trees, with foliage blown off trees and large trees blown down; caravans and poorly constructed signs destroyed; low-lying escape routes cut by rising water 3–5 hours before arrival of hurricane centre; land continuously lower than 1.5 metres (5 feet) may be flooded up to 13 km (8 miles) inland.				
4 (extreme)	920–944	59–69	131–155	4.0–5.5
Some complete roof structure failures on small homes; shrubs, trees and all signs are destroyed; complete destruction of mobile homes and extensive damage to doors and windows; major damage to lower floors of structures near shoreline; land lower than 3.0 metres (10 feet) may be flooded and will require massive evacuation as far inland as 10 km (6 miles).				
5 (catastrophic)	less than 920	more than 69	more than 155	more than 5.5
Complete roof failure on many homes and industrial buildings; some total building failures and complete destruction of mobile homes; all shrubs, trees and signs destroyed; major damage to lower floors of all structures located less than 4.6 metres (15 feet) above sea level and within 150 metres (500 feet) of the shore; massive evacuation from low ground within 8–16 km (5–10 miles) of the shore may be required.				

We know, for example, that these systems are not observed at all in the South Atlantic whereas they are really very common in the north-west Pacific. Table 2.3 offers some statistics related to the frequency of named tropical cyclones for a 20-year period.

On average, therefore, one-third of the world's tropical cyclones occur over the north-west Pacific Ocean – as typhoons. Even in the least active year with 17 systems, this region still takes third place globally, based on the highest annual totals in different ocean basins.

Figure 2.1 shows the activity in the north-west Pacific for a more

Table 2.2 The Australian Cyclone Intensity Scale

Category	Strongest gust (metres/second)	Typical impact
1	<35	Negligible house damage; some to crops, trees and caravans; craft may drag moorings.
2	35–47	Minor house damage; significant damage to signs, trees and caravans; heavy damage to some crops; risk of power failure; small craft break mooring.
3	47–63	Some roof and structural damage; some caravans destroyed; power failures likely.
4	63–78	Significant roofing loss and structural damage; many caravans destroyed and blown away; dangerous airborne debris; widespread power failures.
5	>78	Extremely dangerous with widespread destruction.

Table 2.3 Frequency of tropical cyclones by region

Ocean	Annual mean	Highest annual total	Lowest annual total
North-west Pacific	26	39	17
North-east Pacific	13	20	6
Australia	10	17	5
North-west Atlantic	9	14	4
South Indian	8	13	4
North Indian	6	9	4
South Pacific	6	10	2

recent 11-year period in terms of the total yearly count of tropical cyclones and the division into intensity categories. There is variation in frequency from year to year, with a high of 43 in 1996 and a low of 27 in both 1988 and 1998.

The region we in Europe hear most about – the north-west Atlantic including the Caribbean – comes only fourth, with only about one-third the total seen typically in the typhoon region. The highest and lowest totals remind us that there is a lot of natural year-to-year variability in their

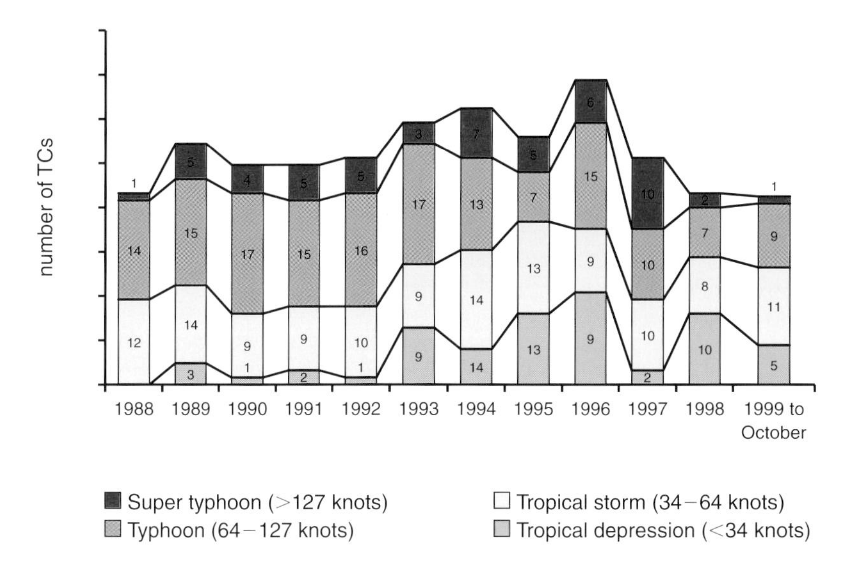

Figure 2.1 Tropical cyclone (TC) activity in the North-West Pacific basin (1988–99).

number. On a coarser scale, the northern hemisphere sees a typical annual total of 55 while the more watery southern hemisphere, perhaps surprisingly, sees a much smaller total of 25 – surprising because these systems form over the open oceans. One reason for the difference is that the sea surface temperature values south of the equator are quite low, particularly over the south-east Pacific and South Atlantic; cool water stifles the genesis of these disturbances.

In addition to these geographical clues, the strong seasonality (Table 2.4) in tropical cyclone formation points again to the role of warm ocean water.

The numbers in Table 2.4 represent (to the nearest whole named tropical cyclone) the monthly average total for each hemisphere. There is a clear

Table 2.4 Mean monthly frequency of tropical cyclones

	Jan	Feb	Mar	Apr	May	Jun	Jul	Aug	Sep	Oct	Nov	Dec
N. Hemisphere	1	0	0	1	3	5	9	11	12	8	4	2
S. Hemisphere	7	6	5	2	1	0	0	0	0	0	2	4

indication of the maximum occurring in August and September in the northern hemisphere and in January and February in the southern. This shows once again the importance of warm water since these are the periods when sea surface temperatures are at their highest. A comparison of 11 recent tropical cyclone years in both the northern and southern hemispheres corroborates these differences. For all tropical cyclones from depression status upwards: the northern hemisphere's most active year during this period was 1992 with 83; its least active was 1991 with 61. By contrast, the southern hemisphere's busiest year was 1996–97 with 39; the least active was 1990–91 with just 17.

Breeding grounds

Much is known about the origin and development of hurricanes and related systems. The advent of weather satellites (the first was launched in 1960) has meant that meteorologists can spot and follow the tell-tale cloud signatures presented by these disturbances. They become organised in such a way that the change in their cloud pattern as seen from space gives forecasters a very useful indication of their likely maximum surface wind speed.

Many of the hurricanes that hit the Caribbean, Central America and the United States are born more than a week earlier over West Africa. That region of the tropics sees many travelling disturbances in the summertime; they track from east to west across the zone of very strong temperature contrast between the baking Sahara to the north and the cooler Gulf of Guinea to the south (Figure 2.2).

Warm sea

Some of these rain-bearing features do not show their true colours until they arrive over the warm waters of the tropical North Atlantic in the vicinity of the Cape Verde Islands or further west. The evaporation of water from the tropical ocean is critical in fuelling the system, allowing it to grow into a more powerful disturbance. Among other factors, the rate of evaporation depends directly on the sea temperature. The water should be at least 27°C at the surface and in the upper few tens of metres. However, this is not the only ingredient in the recipe leading to hurricane strength and not all hurricanes arise off West Africa; some attain this status much further across the Atlantic towards or even in the Caribbean.

Latitude limit

We know that no organised, circular motion occurs within about 5 degrees latitude of the equator. This is related to the way in which the spinning Earth imparts spin to the atmosphere – it is the spin around the local vertical that is

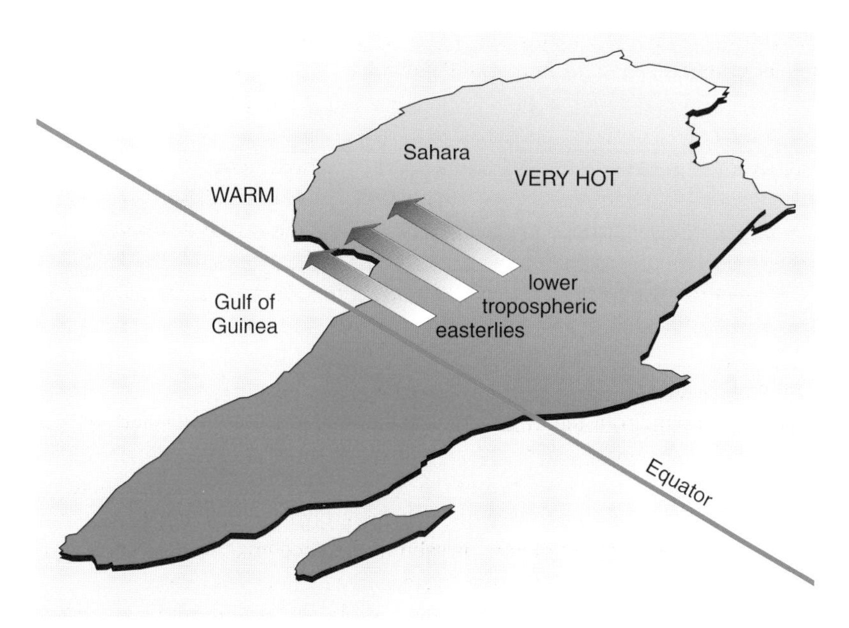

Figure 2.2 The West African low-level temperature gradient during the northern summer. The 'reversed' thermal pattern, in which the temperature increases polewards, is associated with an easterly flow.

of interest. This is strongest at the poles and zero at the equator because, like a top, the Earth spins around an axis precisely at the poles.

Curving flow

In the lower atmosphere, above the warm ocean surface, it is important that there is *cyclonic vorticity* in the flow, a pre-existent degree of 'spin' in the air. One favourable area for this is in the region of the Trade Winds.

Temperature change with altitude

Very deep cumulus clouds are an integral part of hurricanes and typhoons, etc. Their formation requires a temperature profile (change with height) which will promote their birth and growth through the troposphere. This is crucial and means that the *temperature lapse* (the rate at which temperature changes with height) upwards must be quite steep so that warm, moist bubbles of air that arise from the surface are buoyant to great heights within such surroundings that cool rapidly with height.

Moisture at middle altitudes

Even if the bubbles of air do ascend within the troposphere, they will fail to become deep and active if the air a few kilometres up is too dry. They must have a relatively damp troposphere through which to grow successfully – low relative humidity a few kilometres up means serious erosion of convective clouds.

Wind change with height

One more factor critical for the whole hurricane (typhoon etc.) to be able to maintain its distinct identity as it runs across the Earth, is that the enveloping atmospheric 'environment' within which it travels must possess a particular pattern of winds in the upper and lower reaches of the troposphere. If the upper and lower tropospheric surrounding winds differ greatly in speed, the embedded storm system will not strengthen. A small *shear* or difference in wind speed favours development.

Susceptible areas

The combination of these factors means that there are, as we have seen, preferred ocean regions where tropical cyclones form. Once this happens, they very commonly track in a general westward direction as disturbances guided by the larger scale easterly flow of the Trade Winds in which they are often embedded. This in turn means that it is the eastern flanks of tropical and subtropical continents – and the tropical and subtropical islands en route – which mainly suffer the brunt of their destructive force.

Figure 2.3 illustrates the tracks of these systems during a three-year period. Near-equatorial areas are not the most vulnerable to damage; regions towards middle latitudes are much more so. Thus, when the disturbances, embedded in deep south-westerly flow, *recurve* towards the north-east, Japan suffers damage from typhoons on occasion as do the north-eastern states of the US.

The ten most vulnerable American states

The statistics of landfall locations along the Gulf and Atlantic states of the USA (for the period 1900–96) provide something of a 'risk' summary for the different states affected by hurricane strikes (Table 2.5). Of course the size of the state is one important factor; other things being equal, one might expect Texas to experience more strikes than Alabama.

It is perhaps surprising that three of the top ten states include New York and two in New England. Equally surprisingly, Georgia and Virginia are missing from the table. These observations summarise how the tracks followed by the major hurricanes, often around the western flank of the Azores

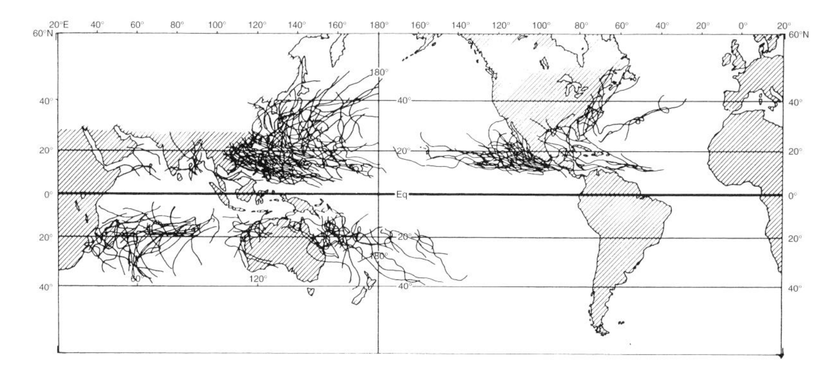

Figure 2.3 Tropical cyclone tracks for a 3-year period.

High, relate to the orientation of the American coastline. The same database for 1900 to 1996 tells us about the risk of a major hurricane (3 or more on the Saffir-Simpson scale) crossing the coastline somewhere between Texas and Maine (i.e. from the Rio Grande to the Canadian border) and for individual states (Table 2.6).

Statistically speaking, Florida is the riskiest state by far – and particularly in September. But this is not a forecast!

Table 2.5 Hurricane strikes in the ten most vulnerable American states (1900–96)

State	**All hurricanes**	**Major hurricanes (3, 4, 5 on Saffir-Simpson scale)**
Florida	57	24
Texas	36	15
Louisiana	25	12
North Carolina	25	11
South Carolina	14	4
Alabama	10	5
New York	9	5
Mississippi	8	6
Connecticut	8	3
Massachusetts	6	2

Table 2.6 Major hurricane landfall in the continental United States (1900–96)

State	June	July	August	September	October
Whole coast	3	3	16	38	8
Florida	0	1	2	15	6
Texas	1	1	7	6	0
Louisiana	2	0	4	5	1
North Carolina	0	0	2	8	1
Mississippi	0	1	1	4	0

Flying into the storms

Information of the current state of a tropical cyclone is obviously of great utility for gauging its intensity at a particular time and for providing data for predicting their future course and strength. When the disturbance is over the ocean, forecasters must rely heavily on the half-hourly images provided round the clock by geosynchronous weather satellites. It is possible to plot their location and, in fact, to estimate their surface wind speed by a careful analysis of the cloud pattern they present to space-based observation platforms.

In addition, there is another critical strand of observations provided routinely for every suspected significant tropical disturbance thought likely to threaten the US and Caribbean region. These are the data logged and transmitted in real time by the flights of 53rd Weather Reconnaissance Squadron based at Biloxi, Mississippi. The squadron have ten C-130 Hercules aircraft at their disposal to monitor tropical disturbances of hurricane and lesser strength. They do so during the official season from June to November inclusive for the Atlantic west of 55° West, the Caribbean and the Gulf of Mexico. The National Hurricane Center tells the flyers where to look.

The first missions are often low-level (150–460 metres or 500–1,500 feet) flights to confirm that an organised cyclonic circulation exists – and to determine its centre if this is so. As a storm intensifies, the flights are made through the system at 5,000 and 10,000 feet (1,525 and 3,050 metres). The elevation of the aircraft's track gradually increases with storm intensification, ending at 40,000–50,000 feet (12,500–15,250 metres) in the all-important outflow region of the hurricane. The eye is penetrated every two hours. These flights are certainly not for the faint-hearted but really do provide critically important data for the operational prediction centre in Miami.

How the damage is done

The fine detail of the typical hurricane, typhoon or cyclone shows us that the really steep horizontal pressure gradient is concentrated across a quite narrow region surrounding the eye. This means, of course, that the strongest winds are concentrated across the same area, wrapped around the eye (Figure 2.4).

At the surface, the lowest pressure is located at the centre of the eye. The difference from this minimum to the edges of the storm is typically 50 millibars, although in extreme cases it can reach 100 millibars. A large proportion of this pressure difference is in fact concentrated across the eyewall region, a narrow zone something like an upright cylinder formed of very deep, vigorous cumulonimbus clouds producing torrentially heavy rainfall. The pressure gradient across this eyewall cloud zone is very steep – it may be only 10 km (6 miles) or so across as a ring that surrounds the outer edge of the eye, typically about 25 km (16 miles) from the centre of the eye. This means that the pressure might drop by quite a few tens of millibars across this narrow band. The eye itself is a roughly circular region right in the centre of the system and is characterised by a very slack horizontal pressure gradient and thus very light winds. The air within the eye sinks in great depth, from the upper all the way down to the lower troposphere which consequently tends to be generally cloud-free or have only scattered, innocuous cloud (Figure 2.5).

The worst conditions seen in hurricanes are hence most often found in the eyewall region, a narrow zone where the winds reach their height and the rain is most intense. The deep convective clouds surrounding the eye are the sign of very warm and moisture-laden air rising rapidly from above the tropical sea surface.

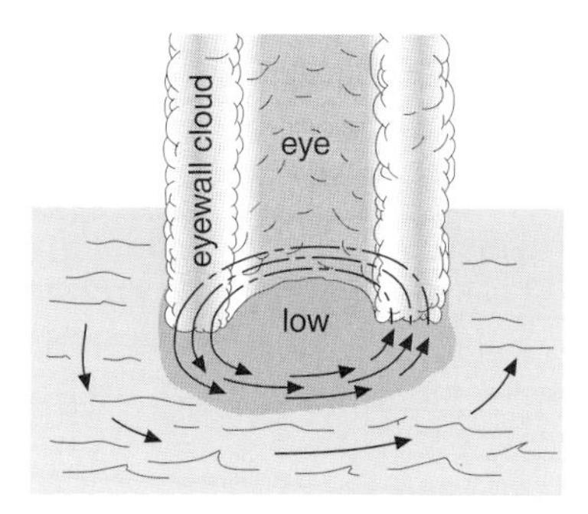

Figure 2.4 The winds at the eye of the storm; the strongest winds are associated with the eyewall's very steep horizontal pressure gradient.

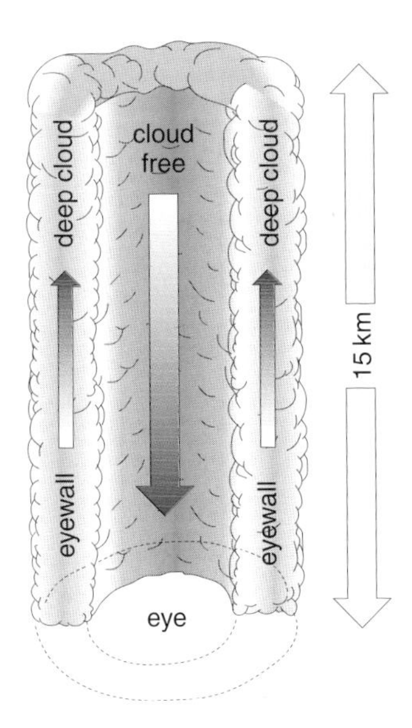

Figure 2.5 The cloud pattern in and around the eye.

The eye is quite a small feature, which means that the intense circulation around it is very tightly constrained. It also means that, when the eye crosses a particular point on the surface, the wind intensifies enormously from one direction, then will drop to near calm, and then pick up again with a vengeance from exactly the opposite direction. Quite how long these changes take depends on the sizes of the eye and eyewall, and how fast the hurricane system is moving. Hurricanes and similar storms move across the Earth's surface typically at about 5 to 10 metres per second (10–20 mph) – the devastating winds are of course blowing within the intense cyclonic circulation (Figure 2.6) and their effects can be devastating.

Spiralling cloud bands

In addition to the eye and eyewall, other characteristic features are the spiral rain bands that wrap partly or sometimes almost completely around the system centre. They display very deep and vigorous convective cloud, producing intensely heavy rainfall, strong gusty downdrafts and the occasional tornado. Both downdrafts and tornadoes may also occur in the eyewall cloud.

Storm surge

Earlier we saw that measuring barometric pressure is equivalent to weighing

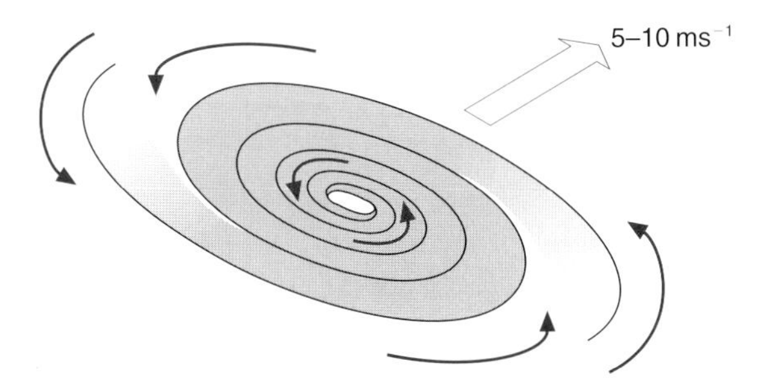

Figure 2.6 The system motion in a tropical cyclone (thick arrow shows system motion, thin arrows are winds blowing around the low in the northern hemisphere).

the atmosphere. This means quite literally that a high pressure region over the ocean will press the sea surface down somewhat in contrast to low pressure, under which the surface will be relatively domed up. This is known as the *inverse barometer effect*: high pressure is linked to a lower sea surface and low pressure to a higher sea surface. The size of this effect is that a change of 1 hPa in the surface barometric pressure is associated with an elevation change in the sea surface of about one centimetre (Figure 2.7). ('hPa' is *hectoPascal,* exactly equivalent to mbar but now in general use to bring it into line with the style of other units.)

For tropical cyclones this means that the normally very low central pressure is 'shadowed' by a domed-up ocean surface known as the *surge*: and the two move in unison. Over the open ocean the surge is typically 50 centimetres (20 inches) or so high but if the system crosses a shallowing ocean area – like much of the US coast from Texas at least round to the Carolinas – it amplifies: the surge deepens significantly as the hurricane moves landwards. Under the most extreme conditions, the surge may exceed 6 metres (20 feet) above normal water levels. The height of and damage inflicted by the surge and its extensive coastal inundation is compounded by the huge waves driven by the very strong surface winds.

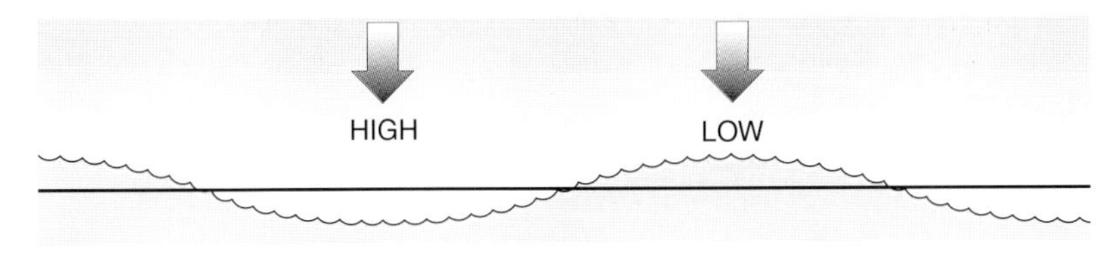

Figure 2.7 The inverse barometer effect: the influence of different mass distributions on the level of the sea surface.

Asymmetric damage

The damage caused by strong tropical cyclones on land is worse on the right of the eye's track in the northern hemisphere (and to the left in the southern). This results from the system itself having a forward speed of, say, 10 metres per second which is superimposed on the powerful cyclonic circulation at the surface. The effect is that the forward right quadrant is normally the worst sector where the wind speeds and the storm surge are most devastating.

Central American devastation in 1998

The 1998 hurricane season was very active. By far the worst was Hurricane Mitch (Figure 2.8) which developed from a lowly tropical depression on 22 October to a very rare Saffir-Simpson category 5 by 26 October. It was without doubt one of the strongest hurricanes ever recorded in the Atlantic Ocean, the Gulf of Mexico or the Caribbean Sea.

Its estimated central sea-level pressure of 905 hPa at 21:00 UTC on 26 October made it the fourth deepest system on record, with sustained surface winds of 80 metres per second and gusts in excess of 89 metres per

Figure 2.8 Hurricane Mitch at 1345 UTC on 26 October 1998. The centre was located east of Honduras with maximum sustained winds estimated to be 155 mph at this time (Courtesy of NOAA).

second (179 and 199 mph, respectively). These wind speeds were sustained for a period of 15 hours, exceeded only by Hurricane Camille in 1969 and Hurricane Dog in 1950. The lowest central pressure value on record was estimated to be 888 hPa in Hurricane Gilbert on 13 September 1988.

Mitch's track

Mitch spelled problems for Jamaica and the Caymans early in its lifetime but really produced its most massive disaster a day or so later. At 21:00 UTC on 27 October it was a category 5 storm off the north coast of Honduras, some 100 km (62 miles) north of Trujilo. Mitch moved slowly over the next couple of days before making a devastating landfall across coastal regions of Honduras. It progressed sluggishly across the Honduran mountains, arriving at the Guatemalan border some two days later – on 31 October (Figure 2.9).

While its winds decreased somewhat as it moved inland, Mitch's slow movement and torrential rainfall meant that up to 60 centimetres (24 inches) per day fell across extensive mountainous areas, leading to horribly massive flooding and mud slides on the steep slopes. Hundreds of thousands of

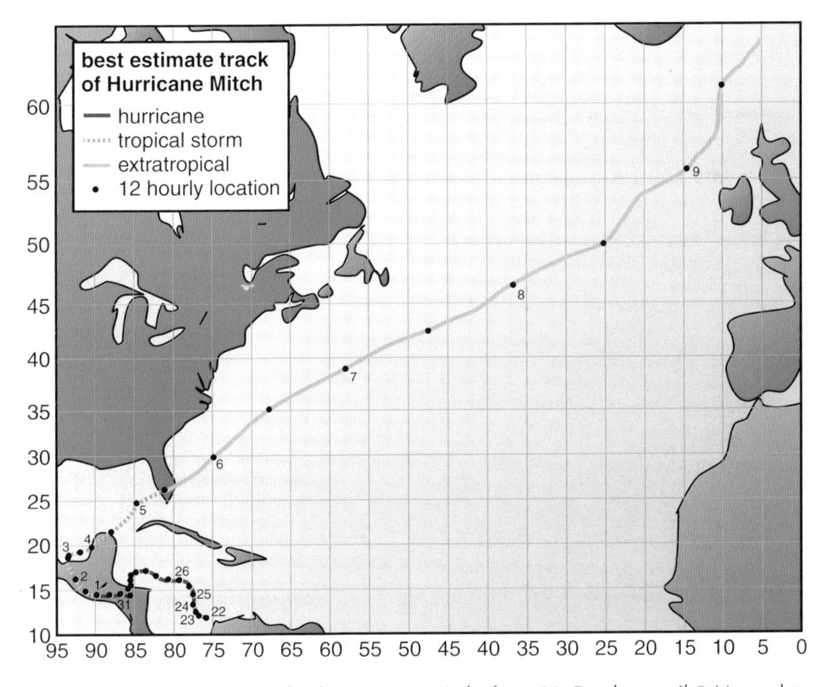

Figure 2.9 Best estimate track of Hurricane Mitch, from 22 October until 9 November 1998.

homes were destroyed. Mitch left a trail of devastation including an estimated 11,000 fatalities, up to 18,000 people missing and damage reckoned to total over $5 billion. The countries worst affected were Honduras, Nicaragua, El Salvador and Guatemala.

On 2 November Mitch left the mainland, heading north-eastwards to run across the western side of the Yucatan peninsula and thence as a reinvigorated tropical storm that rampaged through the Keys and other parts of southern Florida and the Bahamas during 4 and 5 November. After this amazingly long, active life in the tropics it became extratropical on 5 November to become just another travelling middle latitude low pressure area.

Misery in Honduras

Honduras was the worst affected nation although its neighbour Nicaragua suffered almost as badly. In Honduras, about 6,500 people were estimated to have died with a further 11,000 still missing soon after the hurricane. In addition, up to 1,500,000 others were displaced or made homeless: in fact around 20% of the entire population were made homeless.

Helicopters were the only practicable way to take supplies of food, water and medicines into the affected areas but there were few available. This meant it took several days at least to reach many of the more isolated communities where extreme hunger was rife. Malaria and cholera were reported to be spreading. Over half the country's bridges and secondary roads were washed away – it was estimated that between 70% and 80% of Honduran transport infrastructure was destroyed. One third of the buildings in its capital Tegucigalpa were affected by the floods.

Agriculture also suffered massively. Over 70% of the crops were destroyed, including 80% of the bananas. Maize and corn were decimated while much of the country's high value coffee crop became waterlogged in warehouses. Crop losses totalled around $900 million. Coastal areas and offshore islands were severely affected by the high winds and storm surge. The schooner *Fantome* was lost along with its entire crew of 31 after attempting to take shelter on the south side of Guanaja Island.

All in all, it may take as long as 15 to 20 years to rebuild the nation.

Foreign Aid

Soon after the event, the US pledged $80 million for aid to Central America, while Spain and Sweden offered some $100 million each. Many other countries followed suit with smaller but significant donations. Tonnes of food and other supplies were quickly air-lifted into the region by humanitarian organisations.

Extreme rains

Total rainfall amounts for the worst affected areas under Mitch's track could, for the most part, only be estimated since many of the weather stations were destroyed. It is believed that between 1,500 mm and 2,250 mm (59–89 inches) fell during the hurricane's passage over this mountainous area!

Mitch in context

It seems that the scale of devastation, and particularly that of the fatalities, puts Mitch second on the list of worst hurricanes in recorded history. Table 2.7 illustrates what are believed to be the five worst such systems based on estimated fatalities.

The western tropical Pacific has some of the warmest surface ocean water in the world – which is one important reason why the typhoon season is usually much more prolonged than in other ocean basins. Severe events can occur even in the depth of the northern winter.

A *Super Typhoon* is one with maximum sustained winds at the surface of 63 metres per second (141 mph), falling in category 4 on the Australian intensity scale and in the middle of category 4 (extreme hurricane) on the Saffir-Simpson scale. Super typhoon *Paka* struck the island of Guam in the US Pacific Territory on 17 December 1997. Up to 3,000 families lost their homes, with most of the island's properties left without power and water. Its average sustained winds reached 78 metres per second (175 mph); at Anderson Air Force Base on Guam, one gust was recorded at 105.5 metres per second (236 mph). This set a new world surface wind speed record, previously held by a gust of 103.3 metres per second (231 mph) in 1914 on top of Mount Washington in New Hampshire. This Guam record has subsequently been outstripped by radar-measured winds in the Oklahoma tornado

Table 2.7 The most fatal hurricanes

Hurricane	Date	Areas	Deaths
'Great' hurricane	10–16 October 1780	Martinique/Barbados	22,000
Mitch	26 October – 4 November 1998	Central America, Honduras, Nicaragua	>11,000
Unnamed	8 September 1900	Galveston Island, Texas	8,000
Fifi	14–19 September 1974	Honduras	8,000
Unnamed	1–6 September 1930	Dominican Republic	8,000
Flora	30 September – 8 October 1963	Haiti, Cuba	7,200
Unnamed	6 September 1776	Point Petre Bay	>6,000

outbreak of 3 May 1999. US federal aid came quickly after this event, helping to repair and rebuild both private and public property. Supplies flown out of bases near Atlanta, Boston and Tacoma included water purification units, power generators, plastic sheeting, tarpaulins, cots, blankets, portable radios, bottled water, ready-to-eat meals and medicines.

Hurricane Andrew

Hurricane Andrew was the most expensive natural disaster in US history. It was named when it attained tropical storm status during 17 August 1992, some 1,300 km (810 miles) north-east of the mouth of the Amazon. It retained this status over roughly the next five days before achieving hurricane status on 22 August when an eye formed as it lay about 1,000 km (625 miles) east of the main Bahamian Islands. Between 00:00 UTC on 21 August and 18:00 UTC on the 23rd, its pressure minimum had fallen by 92 hPa – to 922 hPa. This rare rate of intensification meant that Andrew had developed from a tropical storm to a system bordering on a category 5 on the Saffir-Simpson scale in only 36 hours!

During 23 August, with Andrew about 100 km (62 miles) east of the north-west Bahamas, a reconnaissance flight at 2,500 metres (8,200 feet) altitude measured winds of 84 metres per second (187 mph) in the eyewall. Its sustained surface winds were estimated at that time to be 67 metres per second (149 mph). After creating havoc in some of the north-west Bahamian islands, Andrew drove on menacingly towards south-east Florida. At The Current, a town on the Bahamian island of Eleuthera, waves and surge combined to reach an elevation of 7 metres (23 feet). The anemometer at Harbour Island stuck at the top of its scale – reading 59 metres per second (132 mph).

Andrew tracked frighteningly towards Miami – with all the massive havoc such a system would create over such a huge urban area. Luckily (for Miami!) the eye made landfall at central Biscayne Bay, crossing the town of Homestead just inland. The closest functioning weather station was at Miami International Airport about 10 km (6 miles) outside of the eyewall.

Andrew's central pressure was estimated by aircraft to be 941 hPa at 04:10 UTC on 24 August over the Grand Bahama Bank. As it moved across the Straits of Florida it deepened dramatically to 936 hPa in the following hour and 35 minutes. At 08:04 UTC, a dropsonde (a package carrying measuring instruments) released from an aircraft implied a sea-level minimum of 932 hPa when Andrew was about 30 km (19 miles) from landfall. The best estimate of the central pressure after landfall, using measurements carefully calibrated from those taken by public volunteers in Homestead, indicate a value of 922 hPa. There was evidence of a smaller, extreme vortex embedded

Figure 2.10 Hurricane Andrew at 20:20 UTC on 25 August 1992 approaching the Louisiana coast for its second US landfall (Courtesy of NOAA/GSFC).

within the eyewall circulation. Hurricane Hugo in 1989 showed evidence of a similar type of small-scale feature. Andrew's minimum pressure at landfall was exceeded only by those of The Labour Day Storm in 1935 (892 hPa) and Hurricane Camille (909 hPa) in 1969.

Andrew's sustained surface wind speed at landfall was probably at a maximum in the northern eyewall where the best estimate was that it reached 62 metres per second (138 mph), gusting to 74 metres per second (165 mph). Aircraft reconnaissance at a flight level of 2,500 metres (8,200 feet) at

08:10 UTC, right in the eyewall 20 km (12 miles) north of the centre, indicated a 10-second flight path average of 80 metres per second (178 mph). This flight-level value is compatible with the surface-sustained winds of 62 metres per second that were likely to have occurred. By contrast, the maximum surface wind speed in the southern eyewall was probably near to 49 metres per second (109 mph). This asymmetry in the wind speed around the eye is typical for systems in the northern hemisphere. As a rule of thumb for westward moving hurricanes and typhoons, the wind in the northern eyewall exceeds that in the southern eyewall by twice the forward speed of the system.

Of course, immensely strong gusts are not confined to the eyewall region – they can occur normally in association with the deep convective cloud comprising the spiral rain bands. One best estimate of such gustiness was the 57 metres per second (128 mph) recorded some distance north of the eyewall. Even the forecasters did not escape – a gust of 70 metres per second (156 mph) was measured on top of the building that houses the National Hurricane Center and the Miami National Weather Service Office. This might very well have been the gust that blew the NWS radome away!

After crossing southern Florida, weakening slightly as it did so, Andrew progressed across the Gulf of Mexico to its second US landfall near Morgan City, Louisiana on 26 August (Figure 2.10). Tornadoes associated with Andrew killed two people in that state where, in addition, the maximum storm rainfall of 303 mm (nearly 12 inches) was recorded at Hammond. After its second landfall, Andrew moved inland and weakened rapidly to tropical storm grade after 10 hours and to tropical depression after an additional 12 hours. This quite common weakening over land is linked both to the system being starved of the water vapour from the tropical seas and to increased *frictional convergence*. This latter is an expression of the greater inward flow of the cyclone's low-level circulation over the rougher land surface: more mass is being transferred into its circulation and the central pressure rises – with a concomitant weakening of the pressure gradient and therefore the wind.

The cost of the damage in southern Dade County in south-east Florida was of the order of $25 billion. Fifteen people perished in Florida, three in the Bahamas and eight in Louisiana. Some 250,000 people were left homeless.

Cyclones down under

Australia is affected by tropical cyclones both around its northern fringe and also along its western and eastern flanks. However, they can and do also penetrate quite far into the interior of the continent, most often affecting Queensland, the Northern Territory and Western Australia. The season occurs mainly during the southern summer so there is a risk of dramatic tropical weather over the Christmas period.

Most of the threatened areas are very sparsely populated so loss of life tends to be limited. But this is not always the case. On Christmas Day 1974 Cyclone Tracy hit Darwin (Figure 2.11), killing 49 people and producing many millions of dollars worth of damage. The damaged buildings were rebuilt to be more resistant in any recurrence.

Much more recently (6–12 December 1998), Tropical Cyclone Thelma turned out to be one of the most intense such systems ever observed off the coast of Australia. It grew in the Arafura Sea to the north of Northern Territory and developed quickly into a category 5 feature. It moved towards the south-west across the Timor Sea before making landfall in the Kimberley

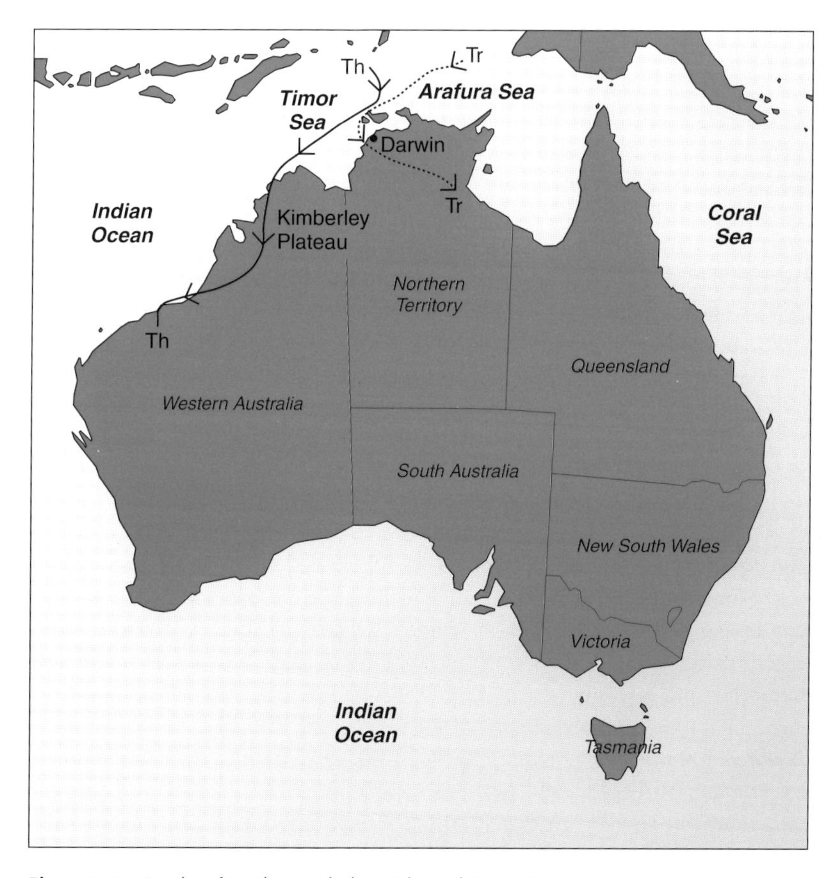

Figure 2.11 Tracks of Cyclones Thelma (Th) and Tracy (Tr).

region of the northern part of West Australia. The damage caused to communities was minor, partly because of the sparse population. On its way to Kimberley, Thelma produced a record 48-hour rainfall of 432 mm (17 inches) at Darwin Airport. The port of Darwin was closed during its passage, schools shut for the day and cyclone shelters were opened for those locals who felt insecure in their own properties. Extensive damage to vegetation occurred to the exposed parts of Melville Island north of Darwin.

Naming convention

The names used for tropical storms and hurricanes in the North Atlantic run sequentially from A to W, and are recycled every six years so those for the 2000 season for example will be used again (perhaps with a few changes) in 2006. If a hurricane is particularly devastating, its name is never used again – perhaps to avoid tempting fate! This has happened in recent years with *Fay* replacing *Fran* (a category 4 system in 1996), *Gaston* for *Georges* (category 4 in 1998) and *Matthew* for *Mitch* (category 5 in 1998).

A similar six-yearly recycling (from A to Z) is used in the Eastern North Pacific. The full alphabet is not exceeded in a season in either of these two ocean basins but the typhoon region of the Western North Pacific does quite commonly experience more than 26 storms in a single season. This region therefore had A to Z lists, used one after the other as necessary: typhoons *Ann* to *Zane* would be followed by a second list running from *Abel* to *Zita*. If the last system of the year is *Rosie*, the first of the subsequent year will be *Scott*. The lists were replaced in 1999 by names contributed by nations in the region. These are not alphabetically ordered – for example, first on one particular list is the Cambodian *Damrey*, second is the Chinese *Longwang*, third is the Korean *Kirogi* and fourth the Hong Kong Chinese *Kai-Tak*.

In Australia, the Weather Bureau has three separate lists of alphabetically-ordered names for northern, western and eastern Australia. As with typhoons, the last cyclone of one season – say *Penny* – is followed alphabetically by the first of the next, say *Russel*. The Seychelles Meteorological Service maintains the alphabetical list used for the south-west Indian Ocean season, the same list being used year after year.

Prediction and warning

Nowadays there are several global weather prediction centres all of which should, by definition, generate, develop and show the decay of the kinds of disturbances discussed here in their operational computer-based forecast models. In the UK, for example, there is an operational global weather forecast model at the Meteorological Office as well as one at the European Centre for Medium-Range Weather Forecasts located in Reading.

The National Center for Environmental Prediction near Washington DC also runs operational models to predict the state of the atmosphere globally for a number of days into the future. Broadly similar schemes are run in other national weather services, including Australia and Canada.

Operational forecasters at the National Hurricane Center near Miami take the forecast products from a variety of different prediction models, including some that are in-house. The prudent thing to do is to keep a watchful eye on the 'freshest' prediction from each centre providing weather forecasts – and to assess whether or not they are all telling the same story about a particular hurricane's track and evolution. There is, of course, great skill and experience involved in making an evaluation.

One important problem for weather prediction in the tropics is the relative lack of observations which, it need hardly be said, are a critical factor in the quality of the forecast. We are generally short of data for both surface and upper-air weather data over the tropical oceans and even much of the tropical continental land area is poorly served.

Meteorologists do, however, have frequent, high-quality images from geosynchronous and polar orbiting weather satellites. The former provide virtually live images, in the form of infrared imagery, of the tropical region every half an hour throughout the day and night. A skilled analyst will be able to pinpoint the tell-tale cloud signature of tropical cyclones and, indeed, can produce a very useful estimate of the likely maximum sustained surface wind speed just on the basis of the image. Such information is invaluable because a reliable estimate of a system's vigour, when it is way out over the ocean far away from the routine observing network, is difficult to come by. The satellite images provide good estimates of the location of a system's centre so that forecasters can use them over time to plot a track and also to check whether the weather prediction models are accurate with regard to a particular hurricane or typhoon's speed and direction.

When a tropical cyclone approaches the US coastline or those of the Caribbean islands and Central America, forecasters will issue two specific types of information:

- *Hurricane Watch* is issued for a specific, well-defined coastal area when there is a threat of hurricane conditions within the following 24 to 36 hours;
- *Hurricane Warning* is more urgent in the sense that it is issued when hurricane conditions are expected within 24 hours.

Action for protection of life and property should begin immediately. They might include:

- staying tuned to radio and television transmission for official bulletins;
- evacuating mobile homes immediately;
- staying at home if the building is sturdy and on high ground (owners should already have checked storm surge history and height above sea level of their home);
- via an official inland route, leaving areas that might be affected by storm surge or river flooding;
- evacuating to an official shelter (individuals should know their locations).

All strong tropical cyclones are monitored particularly well by the geosynchronous weather satellites (Figure 2.12). The forecasts of their track and intensity changes have improved over the years and will continue to do so over the coming decades. More frequent and more widespread observations are one part of this improvement; over some parts of some ocean basins the routine penetration of the systems by special flights is of immense value to better understanding and better prediction.

One perennial problem for the forecasters is, of course, the need to issue timely and accurate warnings. Both *timeliness* and *accuracy* are important; it is very disruptive and expensive for a population to be evacuated, so it is not to be done lightly. It is even more costly, and tragic into the bargain, if

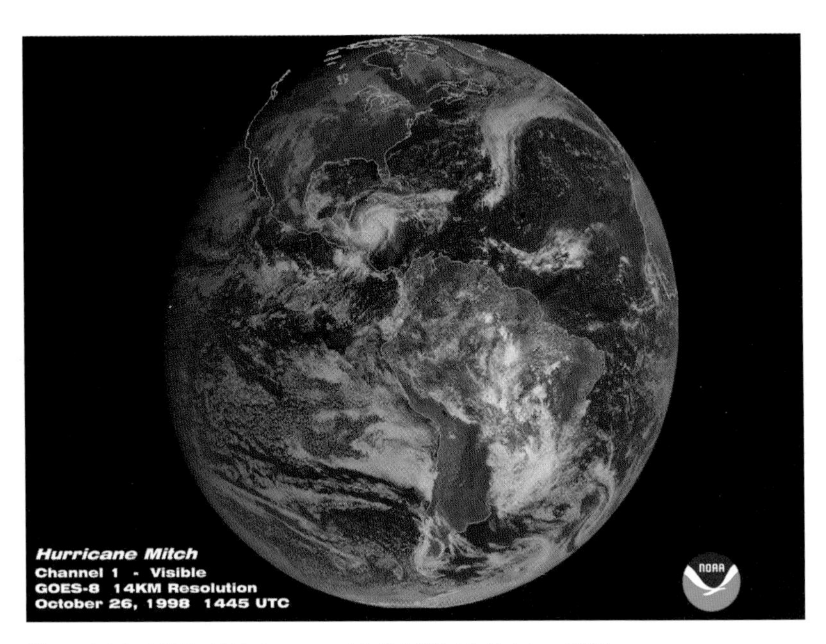

Figure 2.12 GOES visible image at 14:45 UTC, 26 October 1998.

proper warnings are not given so that people die or are injured as well as suffering economic loss. The difficulties for the forecasters are compounded because the most severe damage takes place over a relatively small area; the eye and its surroundings are some tens of kilometres across, so predicting the exact landfall of the eye more than one day ahead is still a great challenge. The forecasting system generally does work – and works well; on occasion, many hundreds of thousands of people are evacuated from low lying coastal areas of the US. If the eye track prediction is significantly inaccurate, the wrong area may be evacuated. The fact that this does not happen bears witness to the generally high quality of the forecast and the forecasters.

On a closing note – the US has the wealth and the organisation to effect massive evacuation and repair if needed. Other susceptible countries, like Bangladesh for example, may benefit from good forecasts but cannot necessarily relay real-time warnings so effectively or evacuate on such a massive scale as more affluent nations.

Hurricanes and *El Niño*

There is a marked year-to-year variability in tropical cyclone activity in the Atlantic and Caribbean, both for tropical storms and major hurricanes (3 and above on the Saffir-Simpson scale). The 1997 season was inactive, with seven tropical storms and one major hurricane during the year. Looking back over the time series of activity, 1983 stands out as a similarly 'poor' year for tropical systems. Could it be that both 1983 and 1997 were somehow influenced by the existence of a very marked *El Niño*? If so, how can a phenomenon in the Pacific Ocean basin play a role in these immense storms across the tropical North Atlantic?

We know there are significant 'teleconnections' within the global atmosphere-ocean system. Teleconnections means that something happening in one part of the word – a large anomaly of sea-surface temperature for example – could influence the circulation, and thus the weather, some thousands of kilometres distant. One way that the tropical cyclone formation region of the North Atlantic and Caribbean can be influenced remotely is by teleconnections in the upper troposphere. This can be linked to enhanced outflow at high levels emanating from very deep tropical cloud. This enhanced outflow is observed as stronger winds which strengthen the westerly jetstream in the upper troposphere and are in turn associated with deeper frontal depressions and therefore unseasonably stronger winds and wetter conditions in areas divorced from the equatorial Pacific.

El Niño is a well-known phenomenon which hits the headlines whenever some dramatic short- or longer-lived weather event occurs, usually in the tropics or lower extratropical latitudes. Perhaps the prime 'signature' of an *El Niño* is the unusual eastward migration of very warm water from the western equatorial Pacific Ocean. Periodically, this means that an unusually warm pool sweeps slowly across the equatorial Pacific over a number of months – until it reaches the South American continent. This warm anomaly is of sufficient magnitude to set off very deep convective clouds – and torrential downpours – where they do not normally occur. The thunderstorms shadow the warm patch as it moves slowly across the ocean.

Large clusters of such convective cloud not only produce localised and often serious flooding in some island groups like the Marshall Islands and the Christmas Islands but are responsible for the upward flow of huge amounts of moist air. The deep clouds are known as 'hot towers' since within their cores they transport very warm, moist air to great heights in the troposphere.

Therein lies one clue to the role of *El Niño* as a source of teleconnections to regions outside the Pacific. The deep convective cloud gathered in one limited region of the equatorial Pacific is the sign of the rising branch of a 'Walker Cell' of which there are a number around the equatorial strip. Such a cell consists of very deep ascent of air along a limited stretch of the equator to produce a westerly or easterly (i.e. parallel to the equator) flow in the upper troposphere above. This air travels some distance east or west respectively before sinking through the troposphere, to return in the reverse direction at low levels.

With an area of anomalously warm water moving slowly across the eastern equatorial Pacific, the ascending branch of one Walker Cell moves in unison with it. There are anomalously strong westerlies in the upper troposphere which penetrate into the Caribbean and tropical North Atlantic. This exported increase in strength at upper levels means that the all-important vertical shear of the wind is increased which is known to be detrimental to the formation of tropical cyclones (see page 24).

It is perhaps therefore not surprising that in a northern summer when *El Niño* is 'raging' in the east Pacific, the chances are that a less active hurricane season might be likely. But there are – as always – some complicating factors! There is some evidence to suggest that when the equatorial Pacific is warmer than average during the *El Niño* events, the tropical North Atlantic is also warmer than average – which would tend to favour an active season. It is believed that the increased ascent over the Pacific is related to increased subsidence over the Atlantic – which would tend to suppress cloud more and lead to sunnier conditions and warmer sea there. On the downside for tropical cyclones, however, increased subsidence tends to mean drier air at middle tropospheric levels which does not favour their genesis.

It seems that the low frequency of tropical storms in 1983 (but not in 1997) was partly related to the *El Niño* increasing wind shear. Increased subsidence related to *El Niño* may, however, have promoted suppression during both seasons in parts of the Atlantic genesis area. There may be yet another factor: during both of these weak years, West African rainfall was below average which may have moderated, through lower soil moisture values, the Easterly Waves that typify the region in the northern summer. They are rain-bearing low-pressure disturbances that run across West Africa towards the Atlantic and can at times develop into tropical storms and hurricanes.

3

Middle latitude storms

Extratropical frontal cyclones

Chapter 2 described a cyclone in general terms as a region of low pressure usually many hundreds of kilometres across; in frontal disturbances so common in mid-latitudes, this dimension usually exceeds 1,000 km (625 miles).

The cyclonic (anticlockwise/clockwise in the Northern/Southern hemispheres) circulation that swirls gradually into low pressure systems at low levels is associated with widespread ascent and oftentimes therefore with widespread cloud and precipitation. These common mid-latitudes features are travelling disturbances that form in regions of sharp horizontal temperature contrast called *fronts* (Figure 3.1). Such thermal patterns occur quite frequently on the western sides of the ocean basins in these latitudes, for example off the north-east coast of the US and the Japan region. The contrasts exist not just at the surface but in depth within the troposphere as shallow, very gently sloping zones.

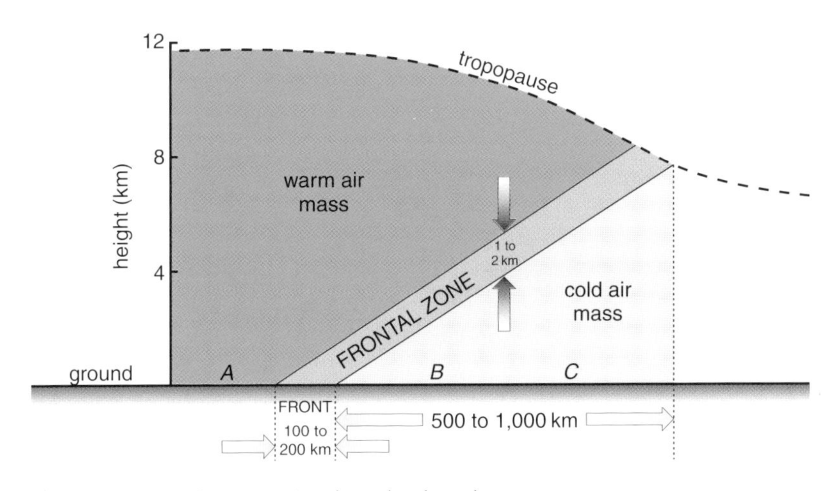

Figure 3.1 Vertical cross-section through a frontal zone.

Once formed, the cyclones (or lows/depressions) tend on average to track north-eastwards across the North Atlantic towards Europe or across the North Pacific towards Alaska and western Canada (Figure 3.2).

Genesis

The formation of frontal cyclones is not simply dependent on a surface thermal contrast but on temperature and humidity contrasts in depth and with key circulation features within the relevant region of the troposphere.

The existence of an elongated stationary front is one ingredient – warm and relatively moist air on its equatorial side and colder, drier air on its polar flank. These two contrasting air masses represent the makings of the warm and cold fronts which evolve in unison with the growing cyclone (Figure 3.3).

As an example, a 'typical' wintertime stationary front might stretch from south-west to north-east off the east coast of the US. This implies that the flow in the upper troposphere will probably also be streaming rapidly from south-west to north-east on the eastern flank of an upper trough; the core of this rapid flow at high levels is the polar front *jet stream*. Although the large-scale upper trough feature will generally evolve and move quite slowly (at a few tens of metres per second), the air in the jet stream may be blasting through the wavy pattern at up to 100 metres per second (around 225 mph) in the winter (Figure 3.4).

The region below the south-west to north-east flow at upper levels is one broadly characterised by convergence, a tendency for the wind at these low levels to swirl into a centre of circulation or towards a long line from either side of it. This is very often related to the fact that the air at high levels is 'divergent', that is, the flow streamlines tend to spread apart. In addition, there are 'short-wave' troughs which are smaller-scale undulations in the flow pattern that actually move through the long wave pattern at upper levels. These important features can also stimulate convergence at low levels and the spin-up of a cyclonic circulation.

Let us assume a short-wave trough has moved over a frontal zone and stimulated a cyclonic circulation. The warm, humid 'tropical' air on the equatorial side of the front will start to move towards the pole (usually north-eastwards in the Northern Hemisphere, south-eastwards in the Southern) while the cold, dry 'polar' air on the poleward side simultaneously starts to move towards the equator.

This counter-clockwise, cyclonic motion forces a pattern in which the warmer, moister air ascends over the colder drier air. Adding water vapour to air decreases its density – and makes it more buoyant, causing the leading edge of the warm moist air to move up and over the cold air as south-

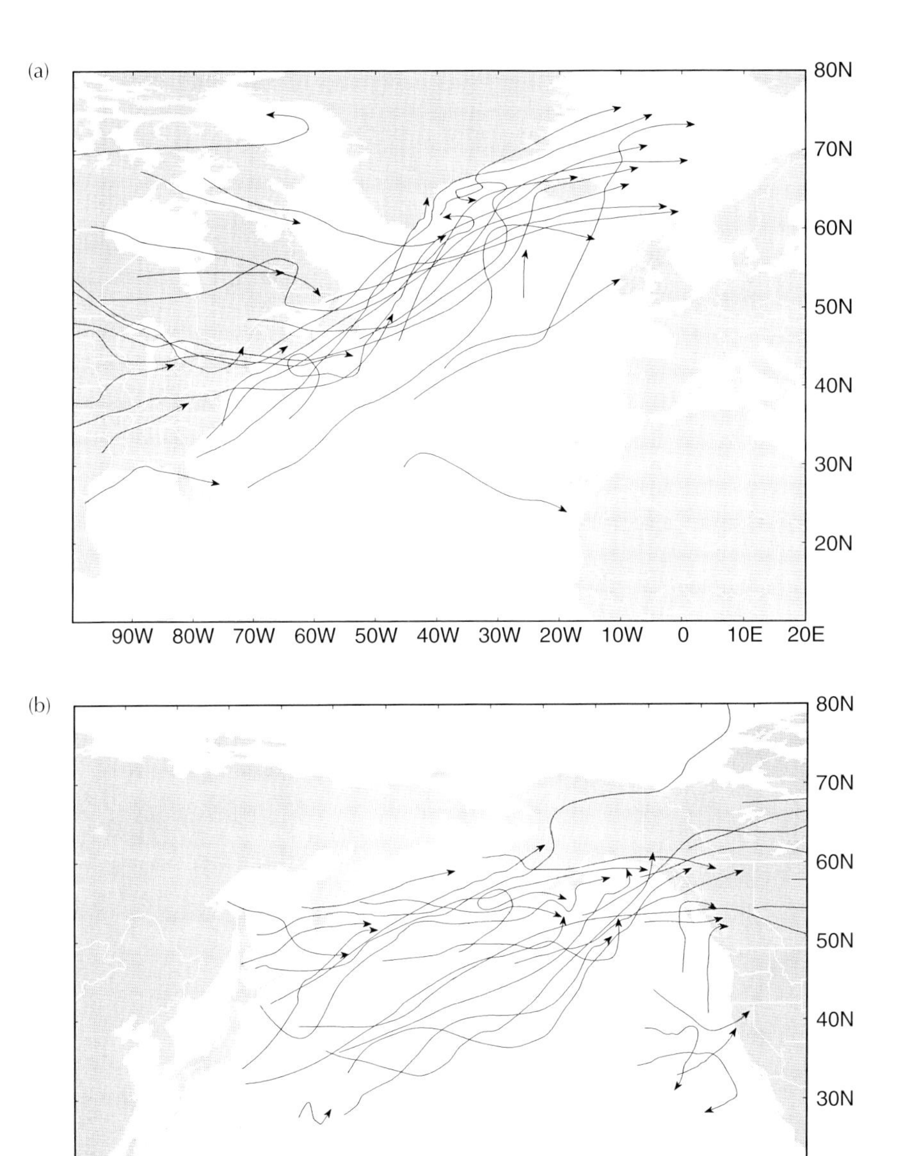

Figure 3.2 Cyclone tracks at sea level in a winter month across (a) the North Atlantic Ocean and (b) North Pacific Ocean.

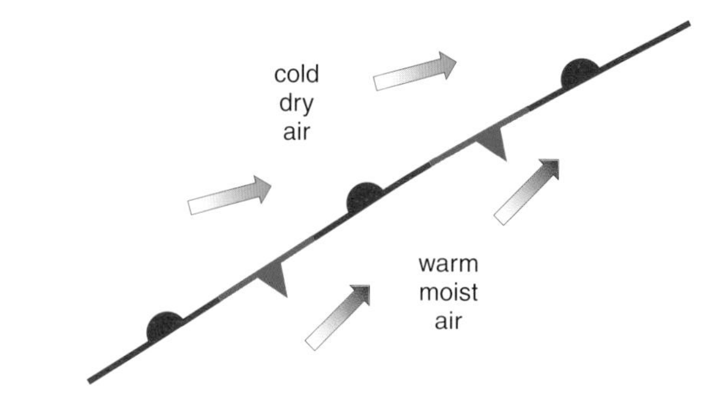

Figure 3.3 Stationary front.

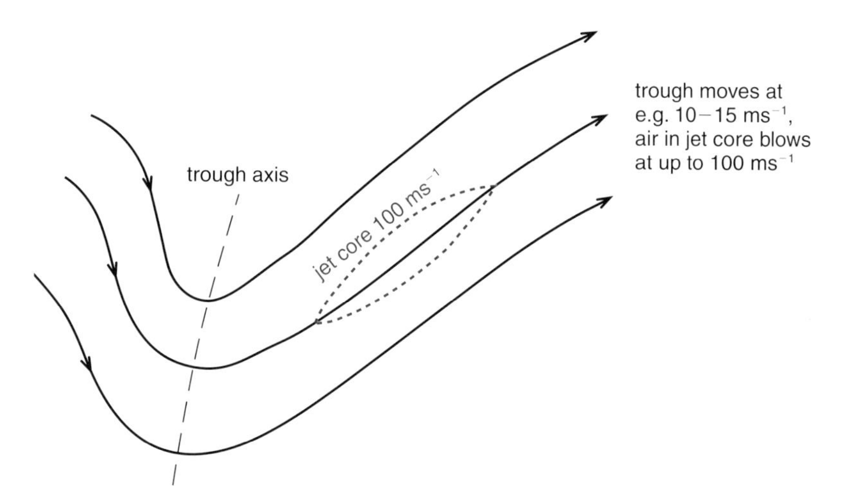

Figure 3.4 Upper trough and jet stream movements

westerly flow in the warm sector. The fact that warmer, less dense air rides over the colder means that the surface pressure will fall. This is because the total mass in the column of such air is less than if it were composed only of cold air.

The outbreak of cold, dry air behind the cyclone undercuts the tropical air, the leading edge of this cold air is the cold front.

The massive volume of warm, moist tropical air that ascends over the warm front is rooted in the 'warm conveyor belt' (Figure 3.5). It streams through the low because it moves faster, leading typically to extensive, deep cloud and widespread precipitation ahead of the surface warm front. Between the warm and cold fronts is the warm sector often characterised by widespread layer cloud and generally light or no precipitation. Rain can, however, become more intense just ahead of the surface cold front in association with an elongated band of convective cloud.

As the cold front passes, the sky tends to become clearer but conditions often change after a short period to scattered showers in the cold air streaming behind the front. This is particularly true of frontal cyclones over the oceans rather than the continents (Figure 3.5).

Energy source

The contrasting air masses lying on either side of the front provide the local source of energy for the storm (ultimately, of course, all these movements are driven by energy from the sun). The gradual ascent of the warmer, moister air over the undercutting colder, drier air leads to the generation of kinetic energy (see box *Energy*) which we sense as wind. In addition, the massive condensation of some of the water vapour in the ascending tropical air gives rise to localised warming due to the release of latent heat (see box *Latent Heat*) in the condensation process. This also influences the energetics of frontal cyclones.

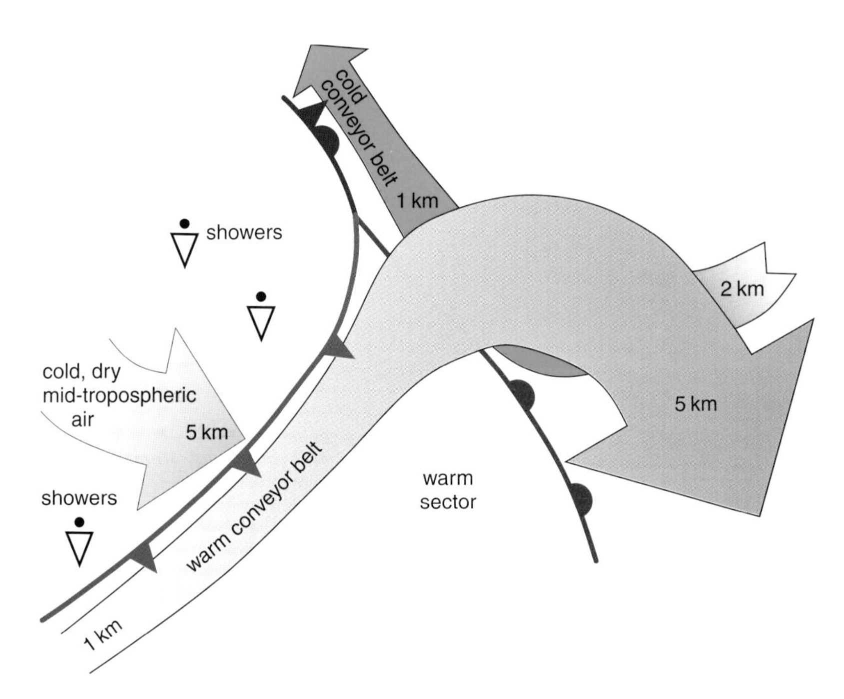

Figure 3.5 A typical depression's relative flow currents.

Energy

Energy is defined formally as 'a capacity for doing work' (so when you wake up in the morning lacking 'energy' and not feeling like going to the office, you are using the word quite correctly!). There are two main sorts of mechanical energy:

- *kinetic energy* is the energy of movement: something travelling at speed (any speed) is able to do things (for instance, a car hitting a lamp standard and knocking it over) which it could not do were it stationary;
- *potential energy* is the energy of position: a heavy weight suspended up high *could* do something as it fell which it could not if it were already on the ground.

In the example of the rising warm air, it is the fact that it is rising and hence is in motion (as a wind blowing) which gives it the ability to do things which would not be possible if the air were at rest.

Latent heat

Imagine a block of ice gently being heated a little at a time. Each additional 'packet' of heat raises its temperature a bit until it just gets to 0°C. Now the temperature does not rise but the ice begins to melt. Assuming time is allowed for the temperature to equalise over the whole block, no rise of temperature will be seen with more heat input until the whole block had melted; then the temperature starts to rise again. The heat which goes in during the melting process is called *the latent heat of melting*. Proceeding in the other direction, water cooling towards freezing does not actually freeze until it has lost *the latent heat of cooling*.

Something similar happens as water is heated towards its boiling point; at just 100°C it does not turn to steam unless a bit more heat provides the *latent heat of evaporation*.

The reasons are that melting ice requires energy to disrupt its crystal structure and allow the freer (but not totally free) arrangement of molecules in liquid water; on the way down, energy has to be lost to allow the ice crystals to reform. Similarly, for water to evaporate, molecules have to break free of the liquid water structure and that, too, demands energy.

Evolution

Once the development of the frontal wave occurs, the whole system tends to follow a classic pattern of evolution (Figure 3.7):

- incipient low: the young feature that forms on a stationary or slow-moving front;
- mature wave: the system intensifies, with falling pressure and strengthening winds around the low centre coinciding with the junction of the warm and cold fronts. Widespread cloud and precipitation occur;
- occlusion: the cold front moves faster than the warm front so that the surface warm air is lifted off the surface, eliminating the energy source for the cyclone. This process starts at the low centre where the fronts are closest; the 'merged' front is an occluded front (Figure 3.6);
- dissipation: with the energy source cut off, cold air swirls into the low centre (Figure 3.6) to increase the central pressure and weaken the horizontal pressure gradients.

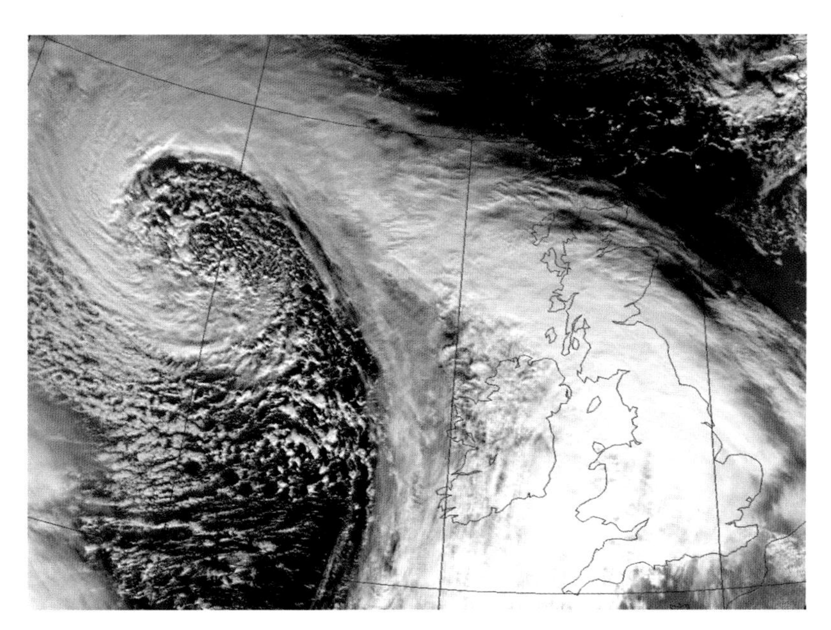

Figure 3.6 Polar orbiting satellite image of a partly occluded frontal depression over the North-East Atlantic at 12:05 UTC, 31 January 2002 (Image copyright NERC Satellite Receiving Station, University of Dundee).

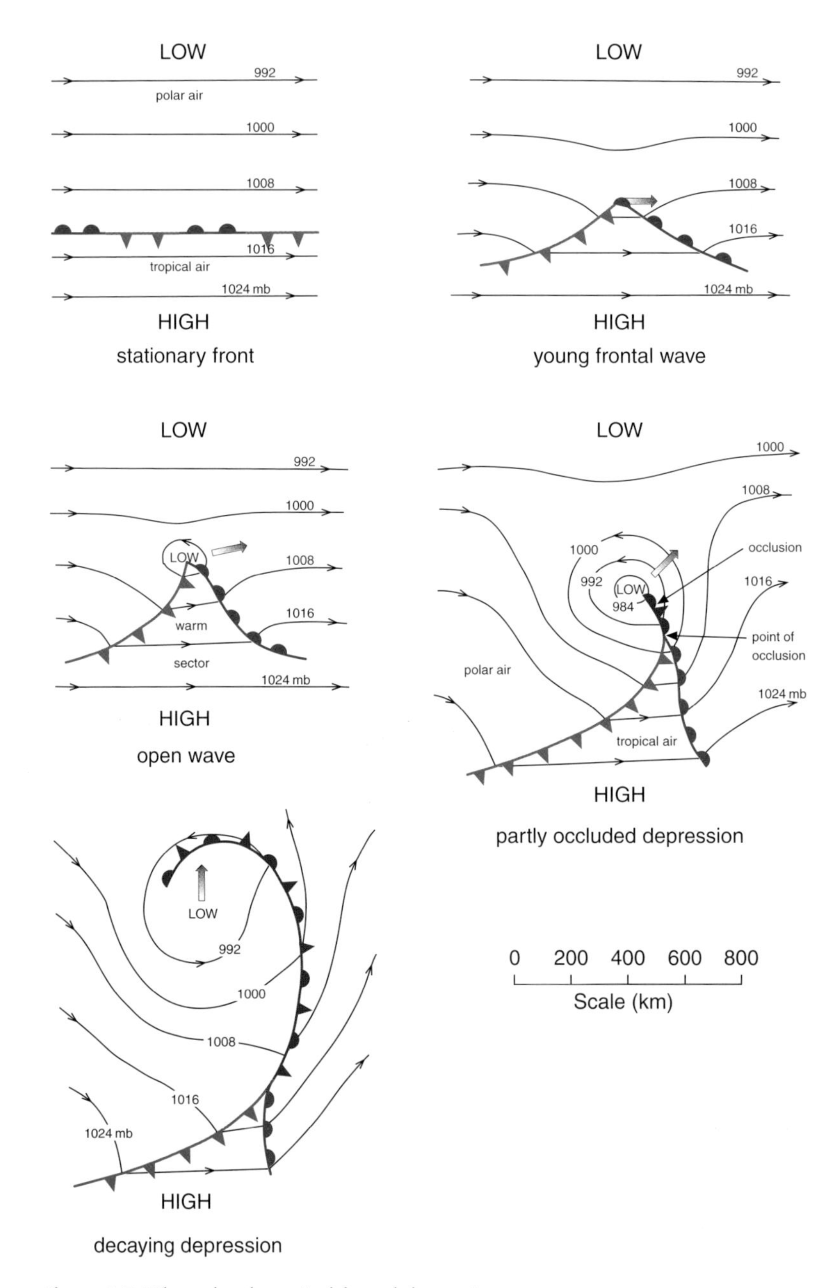

Figure 3.7 Life cycle of a typical frontal depression.

Seasonal contrasts

How powerful the cyclone becomes is dependent partly on the size of the temperature and humidity contrasts across the front with which they are linked. In winter, when the air mass contrasts are much larger, this gradient is much more substantial. This happens because over the North Atlantic, for example, the polar air has swept across a very cold North American continent while the tropical air emanates from an oceanic low latitude source where the temperature is broadly similar to that during the summer season. The result is that winter contrasts are greater than those in other seasons so that the energy available to produce strong winds is larger; that is why gales are much more common in winter than during the summer.

What is the problem?

Extratropical frontal cyclones are not all problems. Outside the tropics they provide many of the world's densely populated areas with water by their widespread, often prolonged precipitation. They are also the principal providers of the frequent winds making possible the generation of wind power.

Wind

On occasion winds can be dramatically damaging (think back to Chapter 2 with its tales of hurricanes and typhoons). The most intense systems produce widespread chaos because of gales or storm force winds. Sometimes their winds are of hurricane force and most certainly there are hurricane force gusts. As with any pressure disturbance, it is the regions affected by steep horizontal pressure gradients which produce the strongest winds; in a deep extratropical low, such areas can be large.

Wind chill

This important phenomenon is related to the physiological impact of cold air on humans and other animals. We need to protect ourselves from significant heat loss during winter by wrapping up in warm clothing. Quite what to wear depends partly on the air temperature and partly on the wind speed. The latter is crucial because strong winds mean that we are exposed to a constant supply of cold air which, if we are not careful, continuously robs us of our body heat. If the air were still we would have a chance of surrounding ourselves with a layer of air which we had ourselves warmed up a little so our own heat loss would be less severe.

Meteorologists express the severity of this phenomenon in a number of ways. One is to combine the observed air temperature with the wind speed to produce an 'equivalent' temperature. This is an expression of what the air

Table 3.1 Wind speed and wind chill

Observed wind speed metres/second	mph	Actual temperature (°C)									
0	0	4	2	0	−2	−4	−6	−8	−10	−12	−14
		Equivalent chill temperature (°C)									
4	9	−2	−4	−7	−9	−11	−14	−16	−19	−21	−23
6	13	−5	−8	−10	−13	−16	−18	−21	−24	−26	−28
8	18	−7	−10	−13	−16	−19	−21	−24	−27	−30	−33
10	22	−9	−12	−16	−18	−21	−24	−27	−30	−33	−36
12	27	−11	−14	−17	−20	−23	−26	−29	−32	−35	−38
14	31	−12	−15	−18	−21	−24	−27	−30	−33	−36	−39

temperature feels like to someone exposed to such conditions; it can be very much lower than the observed value. The 'feel' of the air is also influenced by its relative humidity and whether or not it is sunny. Table 3.1 gives some 'equivalent chill temperatures' – note that for winds stronger than 20 metres per second (45 mph) there is little additional impact. Note that not all severe wind chill events are related to stormy cyclonic features. Very strong winds can and do also blow around the flanks of anticyclones or high pressure systems and these areas can suffer extremely cold conditions including blizzards caused by blowing up lying surface snow.

Another useful way of expressing this effect is to calculate an index relating wind chill to its influence upon outdoor pursuits (Table 3.2) which gives comfort levels (in units of Watts per square metre) as a function of the loss of heat from the body in particular combinations of observed temperature and wind values.

This type of scale is directly useful to the public especially if increasing danger is related to a simple numerical scale from 1 to n. A scale like that, expressing the level of risk as a single digit, is easily understood – just like the Saffir-Simpson scale for hurricanes and the Fujita scale for tornadoes.

Rain and snow

Both are normally widespread and occasionally heavy. A major problem associated with rain and snow is that precipitation totals can be significant if the depression is slow-moving or stationary. Heavy rain can cause localised or more widespread flooding depending partly, of course, on the weather in the days, weeks or months leading up to a particular event; the wetness state of the soil is crucial. If the soil is saturated and conditions are right for flood-

Table 3.2 Wind chill factor (watts per square metre) versus comfort

Wind chill factor	Level of comfort
700	Conditions considered comfortable when dressed for skiing.
1,200	Conditions no longer pleasant for outdoor activities on overcast days.
1,400	Conditions no longer pleasant for outdoor activities on sunny days.
1,600	Freezing of exposed skin begins for most people depending on the degree of activity and the amount of sunshine.
2,300	Conditions for outdoor travel such as walking become dangerous.
	Exposed areas of the face freeze in less than one minute for the average person.
2,700	Exposed flesh will freeze within half a minute for the average person.

ing, a large proportion of any rain will run off straight into the river system, perhaps leading quite rapidly to significant problems.

Widespread and heavy snow produces different problems. Some bad frontal storms may leave 25–50 centimetres (10–20 inches) of snow lying on the ground during the passage of an active system even during one day. This is very relevant for North America where widespread and heavy snowfalls are common each winter. Snow events like this cause serious disruption to traffic and may lead to the isolation of rural communities. Conditions will be even more hazardous if the snow-storm also has strong winds. The forecaster has then to worry about advising of blizzard conditions in which strong winds will reduce visibility to near zero in both falling snow and that whipped up from the surface while power and telephone lines may be brought down, increasing damage and hampering its repair.

In marginal temperature conditions, rain falling onto the ground with a temperature just below freezing can freeze on impact leading to extremely treacherous conditions known as *freezing rain*. In extreme conditions this can lead to major accretions of clear ice on telephone and power lines, tree branches, etc., with serious consequential damage. In the US and Canada such events are *ice storms*; there was a particularly serious one in eastern Canada in the early part of 1999.

Bombs

One important clue to whether a front cyclone will rapidly become intense is to take note of how fast the barometric pressure falls ahead of it. This is routinely done by observers and is reported as the pressure 'tendency', normally the amount of change (in mbar) over the preceding three hours. The reported value (say a fall of 8.0 mbar) is an expression both of the fact that a low pressure system is approaching a station and that it is deepening, i.e. pressures within it are dropping as it moves. A fall of 8.0 mbar in three hours is large but by no means a record. A fall in the central pressure of 24 mbar over one whole day (i.e. an average fall of 1 mbar per hour) in a mid-latitude frontal cyclonic system defines it as a 'bomb'; it is a simple value to measure and warns that the particular system is going to be very windy and possibly very wet or snowy. Such storms can grow from fairly innocuous features to hurricane-force strength within 12–24 hours.

On 15 February 1982 this type of rapid developer produced 45 metre per second (100 mph) winds off the eastern Canadian shore. Massive waves up to 15 metres (50 feet) high led to the disastrous collapse and sinking of the Ocean Ranger oil rig about 300 km (187 miles) south-east of St John's in Newfoundland. All 84 workers perished. The Soviet commercial vessel *Mekhani Tarasov* also sank about 100 km (62 miles) east of the rig; 18 of its 37-man crew drowned.

The fact that these storms breed off or over the eastern seaboard of the US – and do so rapidly – makes them very dangerous to shipping and other maritime operations, and can bury large areas of the densely populated north-east US under very deep snow. They also form outside tropical latitudes over the western Pacific off the Asian Coast, and less frequently over the waters of the central and eastern Atlantic and Pacific Oceans.

'Storm of the century'

The period from 12–14 March 1993 saw the east coast of the US hit by one of the worst storms for many decades. It was remarkable not only for its widespread severity but also for how well it was predicted by the National Weather Service. At 12:00 Eastern Standard Time (EST) on 12 March, the following summary was issued by the National Meteorological Center near Washington DC:

> An extremely powerful winter storm continues to gather strength late this morning south of the Louisiana coast. Latest computer guidance indicates that this storm system will track toward southern Georgia by Saturday morning and deepen rapidly as it lifts northward along the Atlantic Coast during the day. All available guidance at this time suggests this could be a storm of historic proportions along the east coast with potentially record-setting snows over interior portions.

> There are indications that this storm could set all-time low barometric pressure readings at some locations. This intense deepening will result in widespread gale force winds throughout much of the Mid-Atlantic and New England states Saturday and Sunday, with winds possibly reaching hurricane force along the coast. This will likely cause serious problems with coastal flooding and beach erosion. Inland, strong winds will cause blizzard conditions at times with considerable drifting of snow. The combination of the strong winds and heavy snow could also result in widespread tree damage and power line disruption throughout the region.
>
> At this time just where the heaviest snow will occur is still in question. However, the major metropolitan areas from Washington DC through New York and Boston appear to be in the critical path of the storm, and could get significant snowfall amounts prior to a changeover to rain, should it occur.
>
> . . . There is the potential for severe thunderstorms to develop across much of Florida and southern Georgia today and across the south from Florida north-eastward to eastern North Carolina Saturday. The primary threat for severe thunderstorms and tornadoes is expected tonight and early Saturday.

Quite some storm!

Things took off on Friday 12 March with hurricane force winds in Florida – and very high tides there too. Heavy rain fell along the Gulf Coast with snow from Louisiana to Georgia. Deepening of the depression continued throughout the day with winter storm watches and warnings in force for the region stretching from Mississippi right up to Massachusetts. The following is part of the warning issued at 15:00 EST on the Friday by the National Weather Service in Portland, Maine in the far north-east corner of the United States.

> It can't be emphasized enough of the intense magnitude of this storm system. All precautionary action should be completed by everyone no later than early tonight. As a reminder . . . make provisions for adequate food and water along with having battery powered flashlights and radio readily accessible. Travel is not recommended Saturday afternoon through Sunday across the two state area except for extreme emergencies.

The storm did what was expected by the forecasters. In addition to 48 people who perished at sea, 243 other deaths were attributable to the weather. Many were indirect in the sense that people died of heart attacks while trying to clear wet snow from their properties. Others perished in fires, drowned or were trapped in their cars and expired from carbon monoxide poisoning.

Damage was truly widespread – from 5,000 homeless in Reynosa, Mexico hit by a tornado, to three fatalities in Cuba (where Havana's power supply was completely disrupted), to the sinking with the loss of all 33 crew members of the *Gold Bond Conveyor* in 15–20 metre (50–65 foot) seas about 160 km (100 miles) south of the southern tip of Nova Scotia.

Deaths in Florida attributable to tornadoes or other severe conditions ran to 26; a 2.7 metre (9 foot) storm surge hit the state's panhandle region where 15 centimetres (6 inches) of snow fell too! Further north, 33 centimetres of snow at Birmingham, Alabama was a record while 30–60 centimetres (1–2 feet) fell across the southern and central Appalachians. Similar amounts were experienced right up to New England. Locally, falls were very much greater – 127 centimetres (40 inches) was measured at Mount Mitchell, North Carolina, a record-breaking 51 centimetres (20 inches) at Wilkes Barre, Pennsylvania and 109 centimetres (43 inches) at Syracuse, New York. Drifting up to 183 centimetres (6 feet) in parts of Pennsylvania and 152 centimetres (5 feet) in Massachusetts attest to the snow and wind combined. Much further afield, Ottawa witnessed a total accumulation of 139 centimetres (4.5 feet) by the morning of 14 March – the deepest cover in 46 years of records.

Half a million in Georgia were without electricity on the Saturday; elsewhere, some 300 were rescued after being trapped for a day at the Big Walker Mountain tunnel in Virginia by drifts 4.5 metres (15 feet) deep. Every major airport on the east coast was closed at one time or another during the event; Amtrak suspended services on its route between Chicago and the east coast.

High winds wreaked havoc too – either as blizzards whipping up huge waves, or simply for their dynamic force against structures. Some of the highest gusts recorded in the US during the storm are shown in Table 3.3.

As predicted by the National Weather Service, record pressure minima were established for several stations. Over many parts of the South, the values recorded were lower than those associated with the hurricanes affecting the stations over the last three or four decades. The central pressure fell by 58 hPa between 01:00 and 19:00 EST on the Saturday – markedly more than the criterion for the bomb! There were also record low sea-level pressure values during 12–14 March 1993 (Table 3.4).

Total damage to property along the entire coast added up to some $1 billion; snow clearance was estimated at $100 million and overall costs were thought to run up to $4 billion!

Table 3.3 Gust speeds during 12–14 March 1993

Location	Wind speed
Franklin County, Florida	49 m/s (110 mph)
Flattop Mountain, North Carolina	45 m/s (101 mph)
Sth. Timbalier, Louisiana	44 m/s (98 mph)
Sth. Marsh Island, Louisiana	41 m/s (92 mph)
Myrtle Beach, South Carolina	40 m/s (90 mph)
Fire Island, New York	40 m/s (89 mph)
Vero Beach, Florida	37 m/s (83 mph)
Boston, Massachusetts	36 m/s (81 mph)

Table 3.4 Record low mean-sea-level pressure during 12–14 March 1993

Location	Barometric pressure	
	(hPa)	Inches
White Plains, New York	961	28.38
Philadelphia, Pennsylvania	963	28.44
JFK Airport, New York City	963	28.44
Dover, Maine	963	28.44
Wilmington, Delaware	963[1]	28.44
Boston, Massachusetts	965	28.50
Richmond, Virginia	965[2]	28.50

[1]previous low of 972 hPa in 1984
[2]previous low set in Hurricane Hazel in 1954

Avalanche

Widespread snow occurs most often with large-scale ascent within frontal depressions. It is these systems that dump large quantities of snow across the Alps and Rockies, for example. Just what combination of conditions produce avalanches is a rather complex story.

During the winter of 1997–8 there were 26 avalanche fatalities in the US and 20 in Canada – compared with the long-term average of 14 and 6 respectively. Many people perished in the massive avalanches affecting parts of the Alps in the winter of 1998–9, especially in western Austria. About four in every ten people buried in avalanches survive but only three in ten are rescued from complete burial when nothing is visible at the snow surface.

Early in the season, snow can settle onto surfaces that are just marginally above freezing. As subsequent falls accumulate, the lower snow layers are insulated by those above and tend not to be as cold as the very surface where heat can be lost to space very rapidly on clear nights. This change from coldest above to slightly warmer below can lead to a flow of water vapour upwards through the snow layer.

This upward migration can form a layer of 'depth hoar' different from the 'spikey' snowflakes accumulating above. The difference is significant because the depth hoar does not stick together as well as common crystalline snowflakes. This is one potential cause of the massive slipping of the overlying layer as in an avalanche.

A slab avalanche may occur when fresh snow is deposited onto a weak layer. The trigger might be just a small increment of weight from extra snowfall or even that of a single skier or – a more common problem nowadays – a snowmobile. The slab can range from a fairly small breakaway, a few centimetres deep and some 15 metres (50 feet) across moving downhill at about 12–13 metres per second (30 mph), to a devastating one 3 metres (10 feet) deep and 500 metres (a third of a mile) wide careering down the slope faster than 45 metres per second (100 mph).

It is estimated that the US experiences around 100,000 avalanches every year, about 10% of the global total. Daily avalanche outlooks are issued by forecasters working at the National Weather Service Forecast Offices in Denver, Salt Lake City and Seattle.

The riskiest conditions are the very ones that lure many skiers out – a substantial fall of fresh snow after a prolonged 'dry' period. Risks are compounded by complications of slope angle and aspect. These factors mean of course that some areas are real black (or white) spots for being at high risk with 90% of avalanches occurring on slopes at an angle of between 30 and 45°.

Stratospheric plunges

Work over recent years has pointed to stratospheric plunges being one important constituent of many bombs and other rapidly developing extratropical frontal storms. We know that very dry, cold air from the lower stratosphere can become caught up in the circulation of a pre-existent frontal system – this descending air plunges down into the troposphere, in some cases to quite low levels.

The first clue to these blob- or filament-like intrusions can be traced back on a particular type of weather satellite image that portrays the concentration of water vapour about 4–6 km (roughly 13,000–20,000 feet) above the surface. Using these water vapour images, we can trace the movement and evolution of the dry (extremely low humidity) region marking the stratospheric air.

The plunge tends to strengthen the gradients of temperature and humidity within the system and can occasionally be associated with extremely gusty conditions right down at the surface. These important features start their lives as completely independent entities that are often very far from the frontal cyclone, with an origin in higher latitudes.

Storms in Britain

The British Isles do not often suffer extensive and significant damage from extreme storms but there were two in recent times: the storm of October 1987 and the 'Burns' Day Storm' of 1990. Contrary to much popular belief at the time, the former was not a hurricane or even ex-hurricane, but a very deep frontal cyclone tracking across much of Britain.

Strong winds are generally observed more frequently on the northern and western coasts, and higher ground of the British Isles and Ireland. Across the inland areas of central England, even gales are rare events. The required ten-minute average of 15 metres per second (34 mph) for a gale occurs typically no more than twice a year in London compared with gales on some 50 days in Lerwick in the Shetland Isles, north-east of Scotland. Fatalities caused by the very rare, deep frontal depressions are also relatively few – comparable, indeed, to those killed on the nation's roads every day of the year.

The Great Storm

The situation was markedly different three centuries ago, however. November 1703 was generally stormy with the last week particularly so. In the old calendar, it all began late on the afternoon of 26 November in England's West Country with the storm crashing through the south and south-east of the country during the following night. During the previous days the Fleet had ridden out the gales either in harbour or at the eastern end of the English Channel.

The author Daniel Defoe, living at the time in the suburbs of London, was determined to write an account of the storm and its damage. Indeed, the house next door to his had suffered collapse of its chimney during the preceding storm a day or two earlier. Falling chimneys unable to withstand massively strong gusts were the source of much death and devastation. Interestingly, the inquisitive Defoe among others suspected that the storm had run across the North Atlantic all the way from the seaboard of the American colonies.

The October Storm

The storm of 16 October 1987 was foreseen as a deep cyclone which would track across Britain; what was not handled so well by the forecast models was its extreme intensity. It was believed to be the worst storm to cross the region since 1703. There were only 18 fatalities, partly because it ran through during the night and because modern buildings are so substantial.

Some 15 million trees were blown down over southern England, causing massive disruption to transport and the destruction of power lines. The damage-causing gusts reached record values in many places, most significantly along the south coast of England: maxima reached 51 metres per second (114 mph) at Shoreham, Sussex, 48 metres per second (107 mph) at Langdon Bay in Kent and 42 metres per second (94 mph) in central London, all well in excess of the threshold for hurricane-force gusts.

The Burns' Day Storm

This storm was well predicted by the UK Meteorological Office, with warnings to the public issued several days ahead of the event. Unlike the October 1987 depression, it ran through during the day. Although the winds were in general weaker, 47 people lost their lives in Britain during its passage on 25 January 1990. The maximum gust anywhere was at Aberporth in west Wales where 48 metres per second (107 mph) was reported.

In Britain the winter of December 1989 to February 1990 was one of the wettest on record and one of the warmest since 1914–15. Over Scotland, for example, each of the months from January to March 1990 recorded country-average rainfall totals in the top ten monthly rainfalls ever recorded since the series began in 1869! This three-month period gave a Scottish total of 791 mm (31 inches), expected 'statistically' about once in 1,000 years!

The very wet winter meant that water levels by the end of it had no modern precedent; February 1990 was the wettest on record for Great Britain as a whole, with 3–4 times the average over very large areas of England, Scotland and Wales. Before the end of the first week, rivers were in spate across many areas. Heavy rain was boosted by substantial snow melt across northern Britain in the mild conditions. Some Scottish farmland was inundated for several weeks because flood banks had been breached on a large scale.

The dramatic February weather in Scotland was reflected in the flow volume of the River Tay, the UK's largest river in terms of discharge. At its western headwaters at the Lochy power station, 518 mm (20 inches) of rain had fallen in the 25 days up to 4 February. These conditions had led to a completely saturated catchment which, coupled with hardly any remaining storage capacity in reservoirs and lochs, spelled potentially massive flooding problems. Moderate rain after 4 February combined with snow melt led to a peak flow rate of 1,750 cubic metres (385,000 gallons) per second at the Ballathie gauging station on the Tay, the highest flow rate ever recorded in Great Britain.

This whole winter was a dramatic end to a very dry spell as evidenced by the unprecedented water levels in the Chilgrove borehole in West Sussex with records dating back to 1836, believed to be the longest time series of ground water data anywhere in the world.

At the beginning of December 1989, the water levels there were within just a few centimetres of the all-time low during the famous drought of 1976. That soon changed, with an increase in level of 40 metres (131 feet) in eight weeks – a new record for a rate of change. Winterbournes (seasonal streams mainly on chalk) began to flow from spring sources in some areas for the first time since early 1988.

The Braer Storm – a record low

For maritime nations, a serious risk in severe frontal storms is that of oil tankers running aground rupturing their tanks and spilling their cargo. There have, of course, been a number of high profile disasters of this sort, with the *Exxon Valdez* in Alaska perhaps the best known. One well-publicised UK event was that of the tanker *Braer*. She was driven aground in a bad storm in the Shetland Islands very early in January 1993 and stranded for almost a week on the rocks, leaking its massive cargo before being finally completely broken up by the storm of 10 January.

The storm developed the day before on a very strong front – with large thermal contrasts – that was given a 'kick' by an upper trough that passed over it, engaging it in such a way that very strong cyclonic circulation started at low levels. The mass inflow of air was more than balanced by mass outflow aloft by the very strong jet stream there. This led to substantial falls of pressure at the surface, right down to a central value of 914 hPa at 12:00 UTC on 10 January some 400 km (250 miles) south of Iceland (Figure 3.8), representing a drop of 44 mbar during the previous twelve hours, the deepest such system ever recorded in the North Atlantic and possibly the lowest globally apart from the centres of the worst tropical cyclones and of tornadoes.

The intensity of this particular storm was well handled by both the global and the limited area numerical prediction models in the UK Meteorological Office. The quality of the predictions of both the track and the deepening enabled forecasters to issue valuable and accurate warnings as early as 10:00 UTC on 8 January.

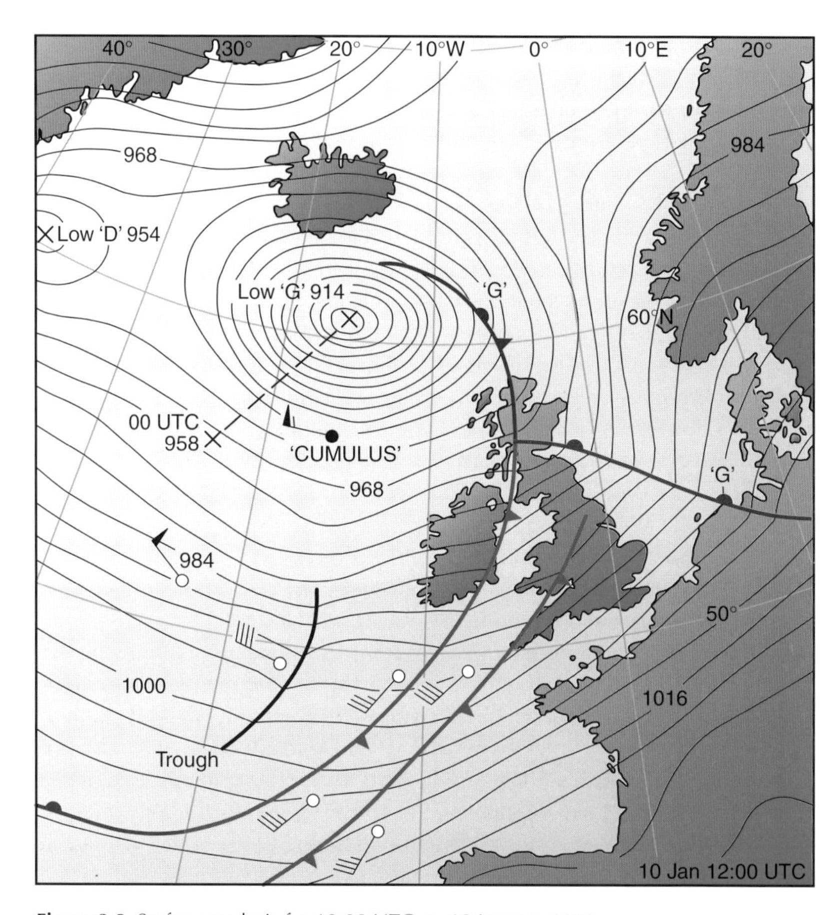

Figure 3.8 Surface analysis for 12:00 UTC on 10 January 1993.

Storm surge

Just over seven weeks later another mid-latitude depression-related problem struck parts of the British Isles. Just as with tropical cyclones travelling across the tropical ocean, deep frontal lows are associated with elevated water levels and strong wind-driven waves. Those living along the low-lying coastal areas of eastern Scotland and England are aware of the occasional risk of flooding from the sea from deep low pressure systems with centres that tend to track eastwards to the north of Scotland and on into Scandinavia.

This motion of deep systems often leads to very strong northerly winds which whip down the North Sea. The UK Meteorological Office houses the

Storm Tide Warning Service which, some years ago, defined a positive surge event as one that led to hourly tidal heights at at least two of the key tide gauges along the East Coast (North Shields, Immingham, Lowestoft, Felixstowe and Sheerness) exceeding predicted levels by at least 0.6 m (2 feet). In practice, this criterion means that about 19 such events occur every year.

Most positive surges affecting the east coast of England tend to show up first as unusually elevated tidal levels at the Stornoway tide gauge sited in the Hebrides, north-west Scotland. They occur in an extensive, strong south-westerly flow in the north-east Atlantic, leading to increased water levels which propagate around the north of Scotland and then down the east coast to the south-east of England. During the Braer Storm for example, the tidal heights in Stornoway were 1 metre above predicted values.

There is a tendency for the surge to amplify as it progresses south through the 'bottleneck effect' of the narrowing of the North Sea so that East Anglia and the Thames Estuary usually suffer the worst. At Wells-next-the-Sea in north Norfolk, the high tide at 06:49 UTC on 21 February was 2.16 metres (7 feet) above the predicted level while at Cromer the surge reached 2.6 metres (9 feet) at 03:00 UTC. This north-facing stretch of the Norfolk coast is susceptible to strong winds from a northerly quarter (Figure 3.9).

The south-eastern sector of England is sinking, possibly producing the occasional collapse of increasingly-undercut cliffs. High water at London Bridge has increased by some 75 centimetres (2.5 feet) over the last century, partly in relation to this gradual sinking (about 30 centimetres or 1 foot per century) and to the gradual increase of sea level. London lies fairly close to the east coast; the centre of the city is potentially open to flooding if heavy rains result in increased flow in the River Thames and its urban tributaries.

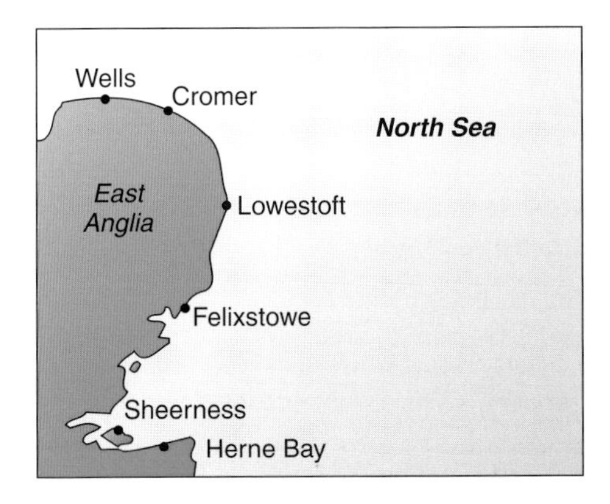

Figure 3.9 East Anglian coast from Norfolk to the Thames Estuary.

This is exacerbated when a North Sea surge is increasing the water level in the river that is tidal right through the city centre and out to Teddington in western suburbia. Flooding in the city would wreak immense damage in the Underground railway system and the basements of the many buildings of the City of London which lie close to the river; the Houses of Parliament also stand on the river bank.

The Thames Barrier at Woolwich was therefore built downstream of most of London to control surging. It is a barrier which can be raised hydraulically to prevent the incursion of extremely high water levels into the main urban area and lowered to allow the passage of shipping. Completed in 1982 after eight years of construction, it does not have to be raised very often (typically once or twice a year) but it was on the occasion of this particular storm. An important stimulus in the construction of the Barrier, costing £1 billion, was the disastrous storm surge of 1953 when 300 people drowned and 65,000 hectares (160,000 acres) of land were flooded across parts of eastern England.

Because of better understanding of the interaction between tide and surge, better weather predictions, and improved modelling of the North Sea itself, forecasts now enable initial warnings to be issued more than a day ahead. They include the length of coastline at risk and the water levels expected to occur at the reference ports noted earlier. Once the warnings are received, the Environment Agency has the responsibility of ensuring round-the-clock staffing of appropriate area and district control rooms, to monitor the situation constantly and to take appropriate action. During this February surge, its Anglian region staff had to secure up to 400 gates and effect other closures in a timely manner. They also ensured that patrols of tidal embankments and susceptible coastal stretches were mounted and liaison was maintained with the emergency services. Over 400 people were evacuated from their homes. The cost of damage to coastal defences was about £2 million, plus an additional 25–50% of that sum in property damage. Significantly, an estimate of the saved costs due to efficient flood management was £19 million.

4

Lightning, thunder, torrent and fire

Thunderstorms

Thunderstorms are relatively small-scale phenomena that are always associated with tall cumulus clouds; very deep convective cloud is classed as *cumulonimbus*. On average in the US for example, a thunderstorm is around 20 km across and lasts some 30 minutes. Thus, as individual weather features, they tend to affect relatively small areas for short periods.

Thunder is, of course, the result of lightning. In an instant, a lightning flash heats the atmosphere locally up to 30,000°C. In the same instant, the intensely heated tube of air rapidly expands and creates the bang we hear as thunder. Near sea level, sound travels in the atmosphere with a speed of roughly 330 metres per second (738 mph). This means that a lapse of three seconds between the lightning flash and the sound of the thunder means the lightning was located about 1 km away (five seconds means one mile away), six seconds means about 2 km and so on. But there is a limit: sound is absorbed so a storm more than 15 km or so (10 miles) distant will not be heard at all, even if the lightning is visible.

At any time there will be very roughly 1,800 thunderstorms active around the world. Many occur within the Inter-Tropical Convergence Zone (ITCZ), some in strong tropical cyclones and some over middle latitude continents in the summertime. Of course, they also happen elsewhere, sometimes over the middle latitude oceans in the wintertime within cold air outbreaks.

Every thunderstorm is dangerous because by their very nature they include lightning which, in fact, kills more people every year in the US than do tornadoes. Of the approximately 100,000 thunderstorms in an average year across the US, about 10% are classed as 'severe'. This term is defined precisely by the US National Weather Service as *a thunderstorm that produces hail that is at least 19 millimetres (¾ inch) in diameter, or a surface wind of at least 25 metres per second (58 mph), or tornadoes.*

Genesis

The atmospheric ingredients needed for the development of deep, electrically active, convective clouds are well understood:

- there must be moisture-rich air to produce clouds and rain;
- the air should be relatively warm at low levels with the temperature falling off rapidly with height;
- this steep temperature 'lapse rate' promotes the growth of bubbles of warm, moist air that rise from the Earth's surface (Figure 4.1);
- the presence of cold air in depth (or its development with the evolution of the circulation) is a critical part of the larger-scale atmospheric environment promoting deep convection in the troposphere.

Dry air at higher levels can also encourage the vertical growth of cumulus. Adding water vapour to a sample of air actually decreases its density; thus, at any given level in the atmosphere, dry air is relatively dense compared with moister air. However, it must not be too dry in depth because that may erode the cumulus cloud as it tries to penetrate through the troposphere.

The notional bubbles of air in such an unstable atmosphere can rise freely but can also be formed if layers of air are lifted by other features like fronts, sea breezes and pre-existent thunderstorm outflows (downdrafts). Moreover, mountains and hills force air to ascend and that, too, may form thunderstorms.

Snappy showers

The first, developing stage of an 'average' shower-producing cumulus cloud

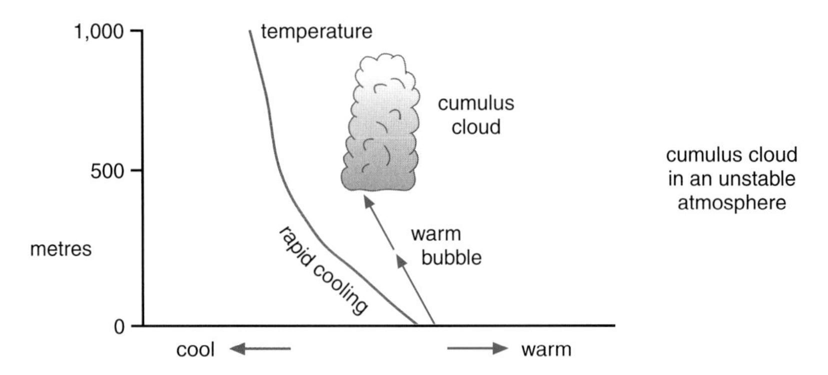

Figure 4.1 An unstable bubble of air.

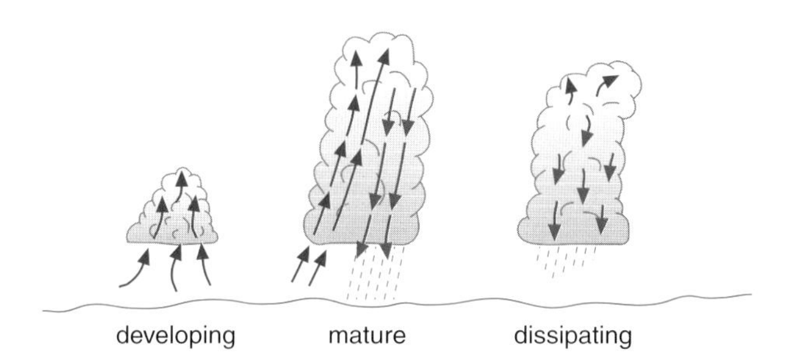

Figure 4.2 Three stages of a shower cloud with an updraft and a downdraft.

lasts about 10 minutes and is characterised by the presence of obvious 'towering' (Figure 4.2). These are brilliant cauliflower-shaped top clouds which grow so rapidly upwards that it is possible to watch the individual sections of the cloud top 'boil up'. The cloud is characterised by updrafts, most of it is warmer than 0°C, with its base up to some 5 km (3 miles) across. At this stage there is usually no rain.

The next, mature, stage lasts on average another 10–20 minutes although it can be a good deal longer. The cloud now has two distinct sections: a troposphere-deep region of convective cloud accompanied by strong updrafts stretching from the surface virtually up to the 'tropopause', the transition between the troposphere and the stratosphere above which there is no fall in temperature with increasing height. Droplets within the cloud grow large enough to start falling back through it as rain. As they fall, the rain drops partly evaporate and this chills the air around them. This in turn creates a downdraft associated with the precipitation, a downdraft strengthened by the drag of the falling rain drops. This growing feature 'counteracts' the ascending moist air which maintains the cloud formation, so such clouds inevitably have short lifetimes. The very fact that they grow deep enough to produce precipitation size drops means they do not last very long.

This type of relatively short-lived system grows and gradually dies in a layer of the atmosphere in which there is no significant change of wind speed with height; they just drift along embedded in the general larger-scale wind. Their 'bubbled' appearance is not distorted by changes of wind speed and direction with height. The clouds generally last for 30 minutes or so, producing a brief rain or snow shower that will leave a trail of precipitation perhaps 10 km (6 miles) long and roughly as wide as the cloud; the length of the swathe depends on how long the cloud remains active and how fast it is travelling.

Airmass thunderstorms

Sometimes the surface heating is stronger than at other times, and the air updrafts can be so deep that bubbling up reaches the tropopause where the clouds flatten out horizontally. The higher a cloud rises, the colder its constituent liquid drops or ice crystals.

The 'freezing level' in a cloud is defined as the location where the air temperature is 0°C. This is, however, something of a misnomer because just about all the cloud droplets are actually still liquid at this temperature. Even at −10°C, only about one 'supercooled' water droplet in a million will have frozen into an ice crystal. The frequency of ice crystals increases as the cloud rises and its temperature falls until, at −40°C, all the water will be frozen to ice.

Making precipitation

The formation and presence of ice crystals is very significant for the process of precipitation. Ice crystals grow at the expense of co-located supercooled water droplets because at subzero temperatures the saturation vapour pressure over ice surfaces is lower than it is over liquid water surfaces. So water evaporates from the supercooled drops (which shrink and disappear) and the vapour is deposited onto the ice crystals (which therefore grow). This is called the *Bergeron-Findeisen process*; it means that growing ice crystals become large enough to start falling through the updraft that helped create them in the first place. As they do so, they collide with supercooled droplets which freeze instantly to enlarge the ice crystals. Some of these tiny crystals will themselves collide and stick together to form snowflakes.

As the flakes begin to fall, they slowly evaporate – as, indeed, do falling raindrops. Evaporation chills the air locally and helps to generate a downdraft. This reaches the surface as a cool or cold, and often blustery, wind which spreads out horizontally, moving way from the cumulonimbus as a gust front (Figure 4.3).

Electric atmosphere

The electrical activity in a deep convective cloud results from charge separation, probably linked to the interaction between the rapidly rising updraft with the ice particles and liquid droplets falling down through it. The more vigorous the up-current, the greater the electrical potential which develops. A typical picture is one in which the upper reaches of a deep convective cloud are charged positively with the lower layers being electrically negative. Once a cloud becomes charged in this way, the surface of the Earth is affected by a positive charge induced directly under the cloud; this charge moves along underneath the cloud as a kind of shadow (Figure 4.4).

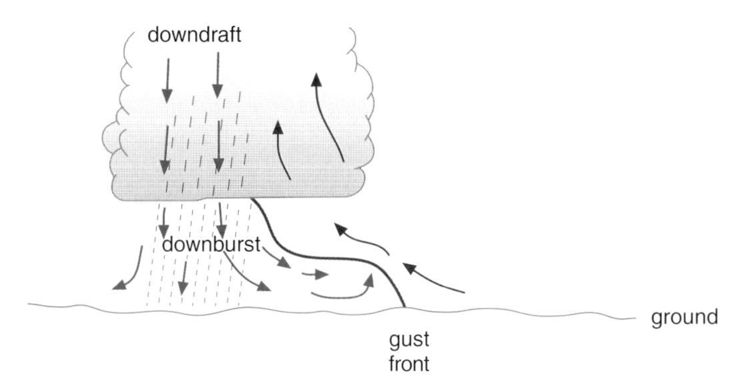

Figure 4.3 Downdrafts in thunderstorms.

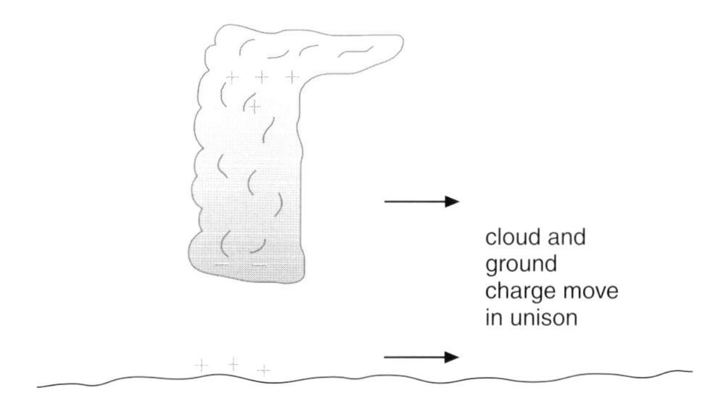

Figure 4.4 Electrical charge below a thunder cloud.

When the electrical potential difference between the cloud base and the Earth's surface is large enough, a lightning stroke takes place. The process starts with a sequence of 'stepped leaders' which are extremely short-lived, short length sections of the stroke that emanate from the cloud, progressing towards the ground at about 50 metres (165 feet) at a time. When the leader comes close to the surface, one or more 'streamers' (short, upward-pointing thin and elongated bright 'tubes' of current) reach upwards from tall buildings, tall trees and other relatively isolated, prominent objects (Figure 4.5). This whole process takes less than a second.

What we see is the brilliant return stroke back up to the cloud – running in a channel that is about 2–5 centimetres (0.8–2 inches) wide, with a peak current of between 10,000 and 200,000 amperes. This may be

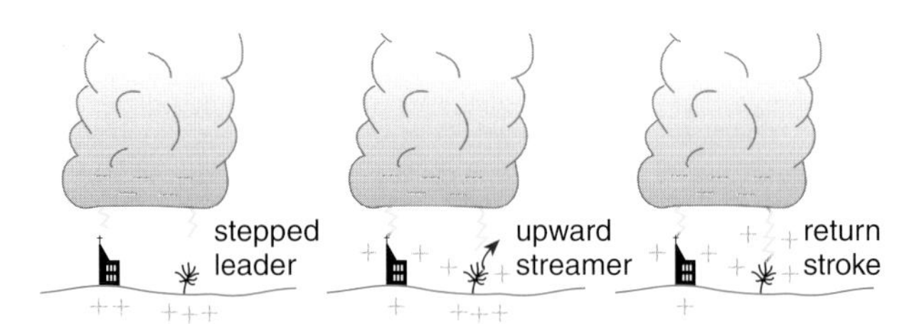

Figure 4.5 The evolution of a cloud to ground lightning stroke.

followed by a sequence of up and down strokes in the same channel: a typical flash has between two and four return strokes.

Electrical activity tends to begin in a cumulus tower when the temperature in its upper reaches falls to between −15°C and −20°C. The thunderstorm reaches its mature phase when precipitation droplets form and may actually be seen falling from the cloud. Because it may well take several minutes for the precipitation to reach the ground, lightning may appear first at this stage. The powerful up- and downdrafts are well developed, with speeds of 25 metres per second (56 mph) or more. At this most active stage in the life of the thunderstorm it is likely to be several kilometres wide and deep. The downdraft spreads across the surface of the ground and tends to cut off the cloud from it sustaining updraft.

Some thunderstorms produce a lot of lightning but no rain (or very little) at the surface. This is often the case in the arid western US. The generally low humidity air drawn up to form the clouds means that cloud base is quite high, with a dry 'subcloud' layer, the layer of dry air lying between the base of the cloud and the ground. Although precipitation does form inside the cumulonimbus cloud, it tends to evaporate before reaching the surface. Many forest fires in the American west are started by these dry thunderstorms (Figure 4.6).

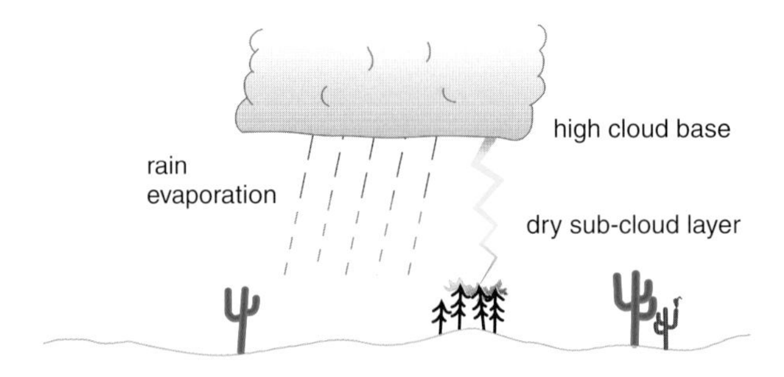

Figure 4.6 Dry sub-cloud layer.

Long-lived severe storms

For convective precipitation to be more long-lived, more widespread and more intense, conditions have to be just right for the larger-scale flow patterns: the wind speed and direction between the lower and upper troposphere must change in such a way that the updraft of warm, moist air is separated from the downdraft which forms as a natural consequence of precipitation generated within the cloud. It is the wind circulation on the larger (synoptic) scale which provides the first ingredient in the formation of a deep, and possibly severe convective, cloud.

The crucial aspect of the wind is that it changes strength with height in such a way that the updraft is not vertical but tilted as it ascends inside the cloud (Figure 4.7). This results in the precipitation growing within the upcurrent falling towards the surface *not back through the updraft* but into the downdraft underlying the slanting supply of warm, moist air. Such a cloud structure permits the updraft to persist for very much longer than the few tens of minutes typical of lesser thunderstorms.

Severe thunderstorms are classified into:

- *multicell* (a 'cell' is one distinct deep convective cloud that has attained cumulonimbus status), in which several of these cells grow at any one time while others are waning. One preferred area for the creation of new cells is the line at which the gust front at the leading edge of its downdraft at the surface converges with other gust fronts. Low-level convergence promotes ascent along a line which may lead to further deep convective cloud; a localised downdraft is known as a *downburst* and, if it is less than 4 km (2.5 miles) across, the term *microburst* is used.
- A *supercell* is an enormous rotating thunderstorm within which the up- and downdrafts are just about in balance so that it can persist for hours. There is a significant risk of tornadoes and damagingly large hail with this type of storm.

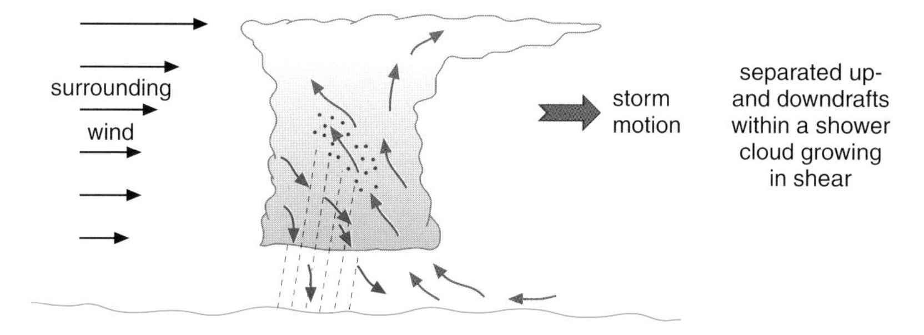

Figure 4.7 A tilted updraft.

Lightning and fire

We often think of lightning as a summer rather than a winter phenomenon and, indeed, the deepest convection is generally associated with the strongest surface heating, which means that there is the perennial region of deep convection strung around the ITCZ.

In the interiors of middle latitude continents, however, heating is seasonal. As a proxy indicator for the peak 'convective' season, the incidence of lightning deaths in the US from 1959 to 1990 shows very obvious maxima in the periods from June to August inclusive. Some 70% of the annual total occurs in the summer with a significant, but much smaller, number in May and September. About 20% of people struck by lightning are killed.

Figures show that the incidence of countryside injuries and fatalities is decreasing with the drift of populations to the cities (Table 4.1). In their place are increasing numbers of incidents related to outdoor recreation: men, mostly aged 20–40, are three to nine times more likely than women to be killed by lightning. Most deaths occur between ten in the morning and seven in the evening, the most popular times for people to be outdoors, especially on sports grounds or in open country. In general, thunderstorms occur during the afternoon and early evening in response to the diurnal heating cycle – just as the golfers are on their way back to the clubhouse! Do not listen to old wives' tales: it is simply not true that 'lightning never strikes the same place twice'. Church spires and towers are fitted with lightning conductors *because* they may very well be struck more than once! In fact, the New York City's Empire State Building is struck many times a year, including one hellish spell of 15 strikes in as many minutes.

On a global basis, lightning strikes the ground about 100 times a second. The US appears first in the world league table for thunderstorms with around 100,000 a year. China runs in second with about 85,000.

Table 4.1 Incidence of fatal US lightning strikes

Period	Average annual number of fatalities
1940–49	337
1950–59	184
1960–69	133
1970–79	98
1980–89	72
1990–99	57

In the US, there are illuminating regional differences in what people are doing when struck. In Colorado for example, with its mountains, many accidents happen during hiking and climbing, whereas in Florida the five top situations were:

- in or near the water;
- under or near a tree;
- near a vehicle;
- near a home or building;
- on a golf course or other sports venue.

Between 1972 and 1991 in the US, more deaths were caused by lightning than by tornadoes and hurricanes. Since then there has been a marked decrease in the average annual total of lightning fatalities probably partly due to more public awareness of the dangers – and better warning of the risk from the forecasters.

Aircraft in flight are susceptible to being struck, usually without damage. Commercial planes are struck once typically every 5,000 to 10,000 flying hours.

The overall totals for the deaths in various American states shows variation; the top seven are given in Table 4.2. If we take into account the area of each state however, we gain a different idea of the incidence of fatal lightning strikes (Table 4.3).

The conclusion is that during this period some of the smaller north-eastern states were more vulnerable than Florida, although the data may be biased in some areas by short-lived and intense bouts of fatal lightning over a limited period. On this basis, Texas is relatively safe with only one fatality for every 583 square km.

Lightning safety

Advice offered by the authorities about what to do when lightning occurs includes:

- if outdoors, go into a building or all-metal vehicle (but not a convertible);
- leave hill tops and open spaces, avoiding wire fences and any conductive elevated objects;
- if no shelter is available, keep well away from tall objects and trees, and crouch down in the open;
- do not use golf clubs, fishing rods, tennis racquets or tools;
- get out of the water, away from beaches and out of small boats/canoes. If caught on open water, crouch down in the middle of the boat, avoiding any metal parts if possible.

Table 4.2 Lightning fatalities in the seven most susceptible states

State	Annual average no. of fatalities (1959–96)
Florida	10
Texas	5
North Carolina	4
New York/Tennessee/Ohio/Louisiana	3

Table 4.3 Incidence of lightning fatalities

State	Square km per fatality (1959–96)
Maryland	36
Delaware	52
New Jersey	57
Florida	61
Rhode Island	116
North Carolina	119
Ohio	124
Tennessee	127
Massachusetts	131
Connecticut	147
New York	147
Louisiana	155

Wildfires

Across the western US, lightning is known to start many serious forest fires. Some begin in areas of very sparse population while others, more newsworthy, impinge very dangerously on large metropolitan areas. These suburban forest fires are becoming an increasing problem not just in the US but also in other countries where expensive suburbs encroach on attractive forest at the edge of expanding cities. Some Australian cities fall into this category. But not all forest fires are caused by lightning; very serious episodes have occurred in recent years in Indonesia where the practise of burning sections of the forest to clear land for agriculture has run out of control, exacerbated by very dry conditions after a period of low rainfall resulting from *El Niño*.

What are the weather conditions that promote forest or bush fires? Lightning can obviously provide the match that lights the fire but rapid spreading depends on a preceding period of very low or zero rainfall, boosted by a dry wind. Together, they will make the vegetation burst into flame like tinder. Strong winds literally fan the flames. The worst fire outbreaks are always associated with windy conditions which aid the frighteningly rapid advance of a forest or bush fire. It may be possible to halt the flames by artificial or even natural fire breaks like a wide strip of pre-burned vegetation, land that has been cleared, or a wide river.

October 1996 saw one of the worst fires to affect southern California in the last decade or so. The region sometimes experiences the hot, dry and blustery Santa Ana wind which blows across the Los Angeles area and its surroundings from the east and north-east. Coming from the very hot interior of southern California, the wind drives fires across small communities and ultimately into the flanks of the huge urban areas in the region. In 1996, the Governor declared a state of emergency in San Diego, Orange and Los Angeles counties.

Some of the worst fires took place in San Diego county where one blaze destroyed some 2,400 hectares (6,000 acres) of forest and 98 homes. Another in the same general area consumed 3,600 hectares (9,000 acres). Further north, a 3 km (2 mile) long fire front threatened Malibu after sweeping across the Santa Monica Mountains. Upwards of 2,500 fire-fighters, assisted by tanker aircraft and helicopters, battled with the flames; six of them, working in Malibu Canyon, were badly burned.

Earlier that year, in late June, more than 6,900 hectares (17,000 acres) in northern Arizona had been destroyed. Hundreds were evacuated from their homes and ten walkers had to be airlifted out of the Grand Canyon because of the danger. Fire fighters from as far away as New Jersey and Florida came to help.

One of the worst events in North America since 17 fire-fighters perished in Yellowstone National Park in 1937 was near Glenwood Springs, Colorado in early July 1994. The fire broke out on July 3rd following a lightning strike and, by the 8th, had been contained to some 20 hectares (50 acres). During that day, however, the wind increased significantly, spreading the fire over 800 hectares (2,000 acres). Estimates of its rate of advance put it at 0.5 metres per second (1.1 mph) which may have cut off some of the fire crews; tragically, 14 fire-fighters died.

In Australia, the worst bush fire was on Ash Wednesday 1983, when 76 people died and over 2,400 houses were destroyed in Victoria and South Australia. In late November 1997, lightning strikes started fires in parts of the same two states as well as in New South Wales. Luckily, most of these were remote and promptly extinguished. Fires in the Blue Mountains, about 50 km (30 miles) north-west of Sydney, were monitored and water-bombed from aircraft. Temperatures in the area at the time reached 40°C (104°F) and winds gusted up to 28 metres per second (60 mph).

Lightning is not of course the sole cause of these fires but studies in the state of Victoria proved it to be the main one there. Table 4.4 shows how lightning is the main cause of forest and bush fires.

Rain

There is a strong link both between the speed and the moisture content of the ascending air in a rain cloud and the intensity of any rain it may produce; the moisture will typically be at a maximum in high summer when evaporation is at its strongest.

The strong ascent in very humid air promotes the growth of huge numbers of large precipitation droplets up to 5 millimetres (0.2 inches) in diameter, compared with small ones at 1 millimetre and an average at 2 millimetres. Assuming they are spherical, the largest drops contain nearly 16 times as much water as the average. Because of the size and number of water droplets they can contain, summertime cumulonimbus clouds may cause torrential rainfall and set short-period rainfall records. If a multicell storm is slow-moving, there may be disastrous local flooding. The highest recorded rainfalls are from such stationary or slow-moving storms, which may also be local regions embedded within larger-scale disturbances like tropical cyclones (Table 4.5).

Table 4.4 Causes of forest and bush fires

Cause	Area burned (hectares)	Per cent of all fires
lightning	53,096	46
public utilities	16,256	14
deliberate	15,649	14
miscellaneous	10,009	9
agricultural	7,799	7
prescribed burns	5,274	5
machinery	2,551	2
campfires	1,446	1
cigarettes	444	<1
unknown	2,974	3

Table 4.5 Some impressive short-term rainfalls

Period	Record	Location	Fall (mm)	Date
24 hours	World	La Reunion Indian Ocean	1,825	January 1966
24 hours	Australian	Bellenden Ker Queensland	960	January 1979
24 hours	British	Martinstown Dorset	279	July 1955
1 hour	World	Shangdi, Inner Mongolia	401	July 1975
1 hour	Australian	Florence Queensland	229	December 1920
1 hour	British	Maidenhead Berkshire	92	July 1901
5 minutes	World	Porta Bello Panama	63	November 1911
5 minutes	Australian	Ingham Queensland	26	March 1977
5 minutes	British	Preston Lancashire	32*	August 1893

*estimated value

Flash floods

Flash flooding after intense rainfall is a major problem in some countries. 'Flash', of course, means that the flood sets in extremely rapidly: within a few minutes or some few hours after a significant downpour. Sometimes the flood affects a community which itself has seen no rain at all; torrential rains may have occurred at some distance from sites quite suddenly inundated by floodwater. Flash floods can move massive boulders, dislodge trees, destroy substantial structures and increase water levels up to 10 metres (33 feet) as well as causing massive and life-threatening mud slides – not uncommon on Caribbean Islands and parts of Central America following torrential rains during hurricanes. In 1998, Hurricane Mitch caused just such flooding in Honduras and Nicaragua.

Between 1972 and 1991, flash floods in the US were the major cause of severe weather-related fatalities, about half of them involved drowning in automobiles – a water depth of 0.6 metres (2 feet) is enough to float most cars and it does not take much water on the move to shift cars quite uncontrollably.

Big Thompson floods

One of the worst flash floods to hit the US was in the canyon of the Big Thompson river which runs down through Estes Park in the Rocky Mountain National Park, and onto the Colorado High Plains near Loveland, a fall of some 760 metres (2,500 feet) in about 40 km (25 miles).

July 31, 1976 saw a celebratory long weekend for Colorado, the 'Centennial State', because it joined the Union in 1876. Some 2,500–3,500 people were staying in or simply driving along the Big Thompson canyon during that day which started off with brilliant weather. Convective cloud started to form over the Rockies in this part of Northern Colorado during the afternoon and, by 6 p.m., thunderstorm clouds had developed in the upper reaches of the Big Thompson catchment. They remained stationary, and very deep convective clouds deposited huge amounts of rain in a short period: estimates gave a total of 250–300 millimetres (10–12 inches), of which 200 millimetres (8 inches) fell in two hours. The change in the river's flow volume was dramatic and very rapid.

At the community of Drake, about halfway down the canyon, the discharge was 3.9 cubic metres (860 gallons) per second before the rain started to fall at 6 p.m. Just three hours later the flow rate had increased to more than 200 times this, much of the canyon's highway (US 34) was washed away. Alerts came late: at 8.30 p.m. a State Patrolman investigating reports of rock- and mud slides radioed that he was escaping quickly, with water up to the doors of his vehicle.

Many who stayed in their cars died; those who scrambled to higher ground sat the deathly night out in safety. At the eastern end of the canyon, where it narrows before issuing onto the High Plains, a wall of water up to 6 metres (20 feet) high rushed down, choked with cars and bits of buildings. In all, 139 people perished, six were never found, 418 homes were destroyed as were 52 businesses; damage totalled $35.5 million. The US Army Corps of Engineers retrieved 197 wrecked cars and over 230,000 cubic metres of debris. In the aftermath, building regulations were modified to require new constructions to be out of the path of any similarly disastrous flood.

Water flowing at a fast walking pace of 1.8 metres a second (4 mph) will exert 323 kilograms on each (vertical) square metre (66 pounds per square foot) of any object it may encounter. Huge volumes of rapidly flowing river water can cause enormous damage. The failure of a poorly maintained and repaired dam, not the weather, caused the worst flash flood in US history; this shows just what can happen. On 31 May 1889 a 22 metre (72 foot) high and 284 metre (932 foot) long dam 40 km (25 miles) above Johnstown, Pennsylvania broke. It was built in 1852, had fallen into disrepair, and had been repaired unsatisfactorily in 1879. A massive rush of water that reached up to 23 metres (75 feet) high at points, crashed down the valley and was still 7 metres (23 feet) high when the unsuspecting community of Johnstown was hit. The official record states that 2,209 citizens perished on that day.

A major British event

A slow-moving depression with embedded deep convective cloud produced huge amounts of rain across the region of Exmoor in North Devon on 15–16 August 1952. On the higher reaches of the moor, upwards of 225 millimetres (9 inches) of rain fell during these two days. Overall, there was a much more extensive area receiving well over 100 millimetres (4 inches) over the same period.

Massive amounts of water ran off Exmoor into the West and East Lyn rivers, which flow steeply down to their confluence in the town of Lynmouth, shortly before issuing into the Bristol Channel. Overnight, 34 people died, 420 lost their homes, 23 buildings were destroyed, 70 seriously damaged, and 130 vehicles were washed out to sea.

Urban flash floods

In many countries with frequent heavy thunderstorms, there is the danger of sudden flooding of major urban areas. The obvious problem with intense rain in towns and cities is that virtually all the rain has to run off the mainly hard, impervious surfaces into the urban drain system. There is no possibility of the 'blotting paper' effect that exists in extensive unsaturated countryside able to soak up quite a significant proportion of a deluge. In Australia, several days in late 1992 saw Adelaide experience flash deluges forced by intense thunderstorms over the Adelaide Hills. Both then and early in 1993, Melbourne suffered similarly. One notorious event in that city in 1972 transformed Elizabeth Street, one of the main thoroughfares, into an angry torrent.

Hail

The size of large hailstones depends partly upon the concentration of cloud water – the higher the concentration, the larger the stones – and whether the hail cycles up and down repeatedly through the cumulonimbus cloud or stays balanced within the updraft. The speed of the updraft provides another factor for growth – the fastest updrafts generate the biggest stones. These conditions are at their optimum for producing hail across the US High Plains during the late spring and summer when supercell storms occur, fuelled by very moist air streaming into the region from the Gulf of Mexico. Hail is quite rare in the Earth's colder regions, where surface heating is not very intense and the water vapour concentrations and cloud liquid water content are both low: for these reasons, any hailstones that do occur tend to be small.

The hailstones travelling up and down in a storm show layering of ice as a sequence of 'onion skins'; those growing only within the updraft have little layering. In the US, the region prone to the largest hailstones stretches from North Texas northwards across the High Plains to the Canadian border. This is also where the long-lived storms are to be found; the many thunderstorms within the ITCZ produce hail but are generally not so intense and long-lived as the multicell and supercell storms of the High Plains. They tend to produce hail on a more moderate scale.

Just as rain falls from travelling cumulus clouds, hail tends to fall along a swathe commonly about 15 km (nearly 10 miles) long and perhaps half as wide. Table 4.6 shows how the speed of the updraft is related to the diameter of the hailstones. The world's largest recorded hailstone fell in Coffeyville, Kansas in September 1970; it weighed in at 757 grams (27 ounces) with a mean diameter of 14 centimetres (more than 5 inches)!

It is not uncommon to hear reports of 'golfball', 'baseball' and 'softball' size hailstones over the Great Plains. Hail this size can seriously damage people, vehicles, property and crops; even pea-size hail can be very unpleasant for exposed individuals. One problem, of course, is that hail can appear quite suddenly and therefore pose a serious threat to anyone in an exposed situation out of doors. The cities of Denver and Calgary suffer occasionally from dramatically damaging hailstorms; both lie just to the east of the Rockies in areas susceptible to large hailstones.

In early July 1998, Calgary suffered millions of dollars worth of damage from a disastrous hail fall, damage which included badly dented cars and waterlogged basements. A little later, on 16 October 1998, Denver was unlucky enough to be hit: so much hail fell that the city's snowploughs had

Table 4.6 Updraft speed versus hailstone diameter

Updraft (m/s)	**Hail diameter (cm)**
10.0	1.2
16.0	1.9
25.0	4.4
45.0	7.6

to clear the highways. Damage to homes was estimated to total $17 million and that to cars $41 million, at a time of year when Denverites are usually breathing a sigh of relief that no severe hailstorms had hit them during the normal peak season from mid-April to mid-August.

A number of dramatic hailstorms have occasionally affected Australian cities. For example, on New Year's Day 1947, hail the size of oranges smashed roof tiles in Sydney, damaged cars and broke the roof of the Central Station concourse; people were injured. More recently, in mid- and late April 1999, the city was hit twice by very damaging hailstorms. The first struck the southern suburbs leaving a trail of damage estimated to cost many millions of dollars. Some of the stones were reported to be half the size of a man's fist, breaking through tile roofs into peoples' living rooms.

Then, a few days later, a similarly sudden severe storm tracked across densely populated areas in the centre of the city, depositing up to grapefruit-size hail onto vehicles, and domestic and business properties. Damage estimates included 20,000 buildings and several tens of thousands of cars. So much serious damage was inflicted to house roofs and windows that thousands of tarpaulins were deployed over houses as a temporary measure. The Insurance Council of Australia estimated that the event was the third most expensive natural disaster in the nation's history after the Darwin cyclone of 1974 and an earthquake in Newcastle, NSW in 1989.

Problems are not limited to the North American and Australian continents. At least twenty-six people were killed in eastern Uttar Pradesh in north-east India in early December 1997. It is not clear whether all the deaths were attributable to being struck by large hailstones or to some other severe-storm related cause like lightning. Widespread damage to property and crops also occurred.

Even in the British Isles hail can wreak quite widespread damage. There is a well documented case of serious damage in August 1846 when dramatic thunderstorms spread overnight through Hampshire, Dorset and Somerset. Temperatures across many parts of eastern England reached 31°C and possibly 32°C in London, hot for these areas but by no means unheard of. Damage was scattered along tracks from south-west to north-east from Berkshire to Kent and Rutland to Norfolk. Not surprisingly, greenhouses suffered serious damage; hailstones estimated to be up to 25 millimetres (1 inch) in diameter smashed thousands of panes of glass in nurseries while many other windows in private and public buildings in the London area were destroyed. The storm reached its most intense phase over the south bank of the Thames: 79 millimetres (3 inches) of rain were recorded in a little over 2 hours during the mid and late afternoon. In some places across the east Midlands, 50 lightning flashes a minute were counted at the height of the storm during the later evening.

Wind

Thunderstorms are very often accompanied by strong gusty conditions at the surface; these can occasionally be very dangerous for aircraft landing or taking off.

The downdraft, an integral component of the circulation of mature systems, spills down to fan out sideways when it reaches the ground or sea surface. When specially intense and localised, this is called the *downburst* or *microburst* (Figure 4.8), depending on its lateral dimension.

Usually, the downdraft tends to propagate away from the moving storm, its leading edge moving across the surface as a kind of 'front'. It can be viewed in this way because it is often the boundary between cooler air which has subsided from middle levels in the thundercloud and the warmer environmental air surrounding the storm; it is also termed a 'gust front' since its sudden passage very often signals a significant change in wind strength, direction and gustiness. In very arid areas like the deserts of North Africa, the cool blustery air behind the front churns up dust and sand to produce the 'haboobs' of that region. They have a very dramatic appearance because the dust storm has a very sharp leading, turbulent edge that may be a few hundred metres deep. It approaches rapidly into air that is clear – until the storm arrives.

We have probably all sensed the gradual increase of a cool wind and waving trees during conditions we know are thundery even if there are no storms obviously close by. This wind is often the sign of the low-level outflow from a thunderstorm.

Because they are more limited laterally, downbursts and microbursts are potentially very dangerous. The wind shear (change in speed and/or direction) accompanying them as they reach the surface and splay out

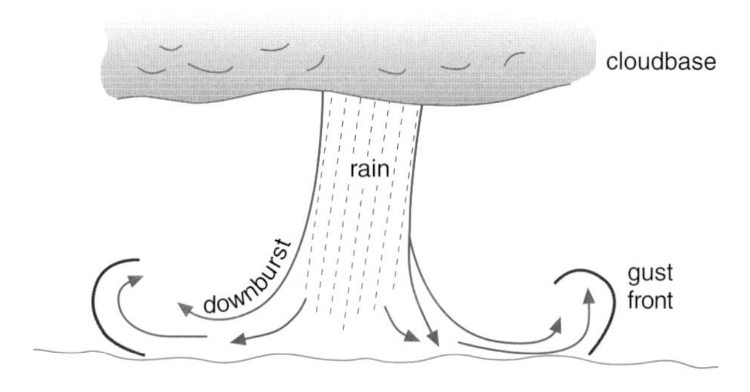

Figure 4.8 A downburst.

appears suddenly and sometimes without visible sign. These conditions are known to have led to very serious commercial aeroplane disasters.

For safety and control, aircraft take off and land into the wind. Flaps are set specifically for both these operations but if for some reason the wind speed and/or direction suddenly changes, disaster is likely to ensue. In the US and other areas susceptible to these severe storms, terminal Doppler radars have been installed at airports to spot the downbursts and microbursts which may sweep across runways from a storm perhaps a few kilometres away. Very short-term warnings can then be issued to aircraft to curtail landings and take-offs until the all clear is advised.

Wind shear weather may also accompany sea breeze circulations, frontal systems and hurricanes and typhoons for example, but the most dramatic events are nevertheless normally associated with severe convection. Severe wind shear is defined as a rapid change in wind direction or wind velocity (direction and speed), causing a change in the airspeed of an aircraft of more than 7.6 metres per second (17 mph) or a variation in the vertical speed of at least 2.5 metres per second (5–6 mph). This may not sound like much but is in fact very significant for an aeroplane taking off or landing with the pilot making a judgement about when to perform an evasive manoeuvre.

An aircraft's airspeed is its speed over the ground, plus the speed of the air (the observed wind speed) through which the plane is flying. Large, wide-bodied jets have ground speeds of around 270 metres per second (around 600 mph) at cruising altitude. This is not normally how fast it is travelling along its flight path because a strong tailwind may increase its speed by up to 30%. An equally strong head wind (normally avoided if at all possible) can cut 30% off the cruising ground speed, meaning that it takes that much longer to arrive at the destination, uses more fuel and may upset schedules for further flights.

The faster the air rushes under and over the specially-shaped aircraft wings, the stronger the lift. A sudden change of wind speed will therefore affect the degree of lift experienced by a plane. This is one of the elements of wind shear and it is therefore particularly dangerous when the speed drops suddenly during take off and landing. Another danger both to the aircraft and to passengers and crew is the existence of Clear Air Turbulence (CAT). Often, very turbulent conditions are associated with obvious features like deep convective cloud – they can be avoided. However, by definition, CAT is not made visible by cloud. It commonly occurs around the flanks of jetstreams where the wind speed changes quickly, moving towards the core. This strong speed shear around an elongated core (perhaps 1,000 km (625 miles) long) is associated with CAT. Its severity – and its location – are well handled by numerical weather prediction models; of course, severe CAT is avoided in the flight plan. 'Air pocket' is an imperfect popular term that relates to the downward turbulent eddies forming part of CAT. If they are strong, the aircraft can fall through some hundreds of metres very rapidly, injuring passengers not strapped in by their seatbelts.

Wet and Dry

Dry microbursts are common in the arid west of the US, where convective cloud base is often high and any precipitation that falls evaporates before reaching the surface. The 'rain shafts' or 'virgae' can be seen evaporating as they descend. This process cools the air within which the rain is evaporating – by evaporative cooling. The cooling leads to a significant downdraft or plunge of cool air from quite great heights (a few thousand metres/feet). The outflow is around 2 to 4 km (6,000 to 12,000 feet) across; the 'ring' apparent when looking down on the system, runs away from the parent convective cloud with a leading edge that can be up to 600 metres (2,000 feet deep). Observations of a number of cases indicate that the maximum speed change across the outflow occurs at a height of about 75 metres (245 feet) above the surface.

While essentially on the same scale as their dry 'cousins', the wet version occurs in thunderstorms that have a much lower cloud base from which rain is falling to the surface (Figure 4.9). Evaporative cooling and drag associated with the falling drops produce the cool plunge in this case. In general, microbursts intensify for the first five minutes after reaching the surface and last a further five to fifteen minutes.

About 5% of US thunderstorms produce microbursts, the majority of which are 'asymmetric' because they are flowing down from a parent thunderstorm that is itself moving. This means that the outflow speed tends on average to be twice as strong in the direction of motion of the parent cloud than on the outflow side that 'trails' the system.

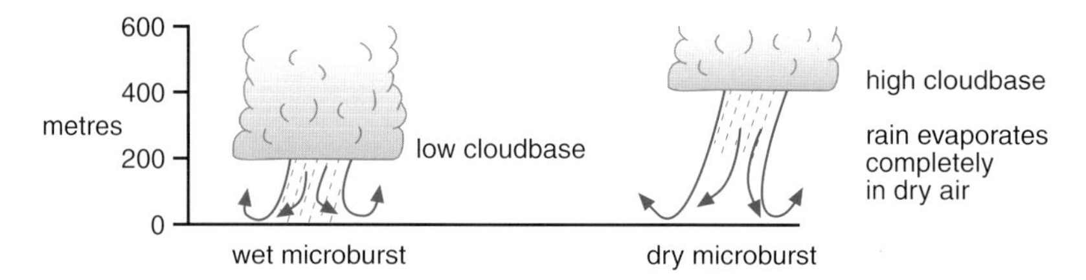

Figure 4.9 Dry and wet microbursts.

Fatal crash

An illustration of the sudden and deadly effect of a microburst on an aircraft in its final landing approach occurred on 2 July 1994. On that day, a DC-9 was coming in to land at Charlotte Airport in North Carolina; the flight deck crew were aware of a few thunderstorm cells in the vicinity of the airport but had been advised of smooth landings by two recently incoming aircraft. Due, apparently, to a radio frequency switch from approach control to the control tower at Charlotte, the flight crew did not hear a warning of wind shear that was to affect the entire airfield.

Rain started to fall some 3 km (2 miles) from touchdown, after which its intensity increased instantly and dramatically. At about 365 metres (1,200 feet), airspeed increased by 4.9 metres per second (about 11 mph) and all forward visibility was lost. The Captain was required to 'go around': to avoid any further problems, the aircraft should circle and try again when conditions improved.

Apparently, a few seconds after zero visibility, the aircraft simply dropped. At the start of the go-around, airspeed was about 72 metres per second (162 mph) with the aircraft at a height of almost 61 metres (200 feet). Engine power increased to begin a climbing right turn. After ascending by some 45 metres (150 feet), airspeed started to decrease. As this continued, the pitch and role attitudes were 15 degrees nose up and 17 degrees right bank respectively.

Attempts to correct to a safe altitude led to a 54 degree nose down inclination as the plane started to descend. A few seconds before the impact, this had changed to 2 degrees nose up but the airspeed was down to 57 metres per second (128 mph). Despite the best efforts of the Captain and First Officer, the aircraft crash landed in these very severe conditions: 37 of the 57 on board were killed.

The accident resulted from attempting a landing in conditions which, just a few minutes earlier, had apparently warranted no undue concern. The type of microburst encountered was probably driven by a rain shaft with a diameter of about 3.5 km (2.2 miles). The event was exacerbated by the probable coincidence of the glide path with a strong burst in the early part of the microburst's life cycle when it would have been at its most intense.

5

Tornadoes

The scourge of the prairies

The strongest winds ever recorded (or in most instances estimated by the severity of the damage caused) are those produced by tornadoes.

In early May 1999, the world's strongest ever measured surface wind speed was sensed in the devastating tornado outbreak that moved through Oklahoma City and some surrounding communities. A mobile Doppler radar on a truck run by a research group at the University of Oklahoma just down the road in Norman measured a surface wind speed of 139 metres a second (312 mph) in the outbreak, beating the previous world record.

Tornadoes are always linked to a cumulonimbus cloud which is itself often one component of a much larger complex of extremely deep, violent convective cloud – the supercell described in Chapter 4 (Figure 5.1). A tornado begins when the distinctive funnel cloud that develops from the base of the thunder cloud makes contact with the ground (Figures 5.2 and 5.3). Contact with the ground is the crucial factor, so long as debris is being thrown up from the surface, there does not even have to be a funnel cloud. 'Touchdown' is the key!

Tornadoes can be associated with a mid-level *mesocyclone* which is a relatively small-scale cyclonic circulation within the parent cloud, not obvious at the surface, but sometimes the precursor of tornado development.

Tornadoes are the most violent of storms, occasionally confused by the public with hurricanes. They are very, very much smaller and very, very much shorter-lived. Hurricanes are typically five, six or seven hundred kilometres in diameter and last for days or sometimes, in one form or another, for up to a couple of weeks (Chapter 2). By contrast, a tornado is typically between 100 and 400 metres (330 and 1,300 feet) across with a damage path stretching along some 2–3 km (less than 2 miles), although the length of the damage swathe can vary from virtually one point to as long as 160 km (100 miles). During its lifetime, often no more than 5 minutes, the width may vary.

It is important to remember that, just like hurricanes, the really intense winds are circulating around the low pressure centre but the tornado as a

Figure 5.1 Supercell storms across part of Oklahoma and Texas at 18:45 local time on 3 May 1999. These were associated with devastating tornadic outbreaks over the region (Courtesy of NOAA/NESDIS).

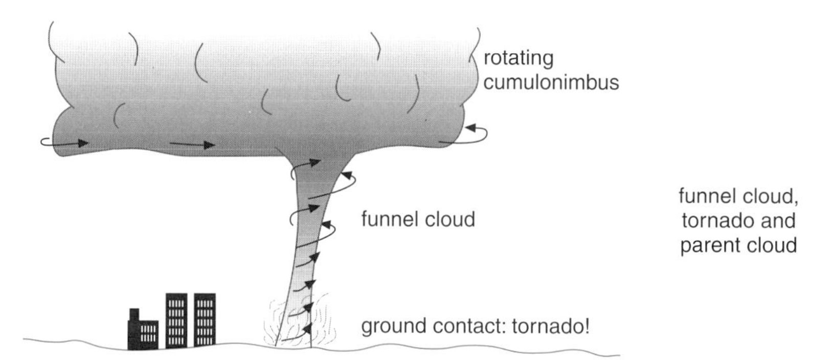

Figure 5.2 Tornado and its parent cloud.

Figure 5.3 Tornado at Dimmit, Texas on 2 June 1995 (Courtesy of NOAA/NWS).

whole moves across the ground surface at a relatively slow pace; although 5–10 metres a second (about 10–20 mph) is common, it can be anything from almost stationary to 25 metres a second (56 mph). Most of them travel from south-west towards the north-east but in certain circumstances their line of travel can be very erratic.

Just like hurricanes and their Saffir-Simpson scale, which is a simple way of conveying severity with one number, tornadoes in the US are similarly categorised by a simple numerical scale (Table 5.1). This was developed by the Japanese-born atmospheric scientist Tetsuya Fujita in the late 1960s on the basis of the type and extent of damage caused by tornadoes.

'Weak' tornadoes (F-0 and F-1) are the most common at 69% of the total. They are responsible for less than 5% of tornado-related deaths and last typically 1–10 minutes. The 'strong' tornadoes (F-2 and F-3) comprise 29% of all cases and generate 30% of tornado-related fatalities; they may persist for 20 minutes or even longer. The 'violent' category (F-4 and F-5) represent only 2% of the total but account for 70% of deaths; sometimes they last more than an hour.

Tornadoes occur in other parts of the world besides the US: in Australia, Africa, Asia, South America and Europe – even in the British Isles! But,

Table 5.1 Fujita tornado severity scale

Scale number	Wind speed metres/second	mph	Designation
F-0	up to 32	70	Gale tornado
Some damage to chimneys; breaks branches off trees; pushes over shallow-rooted trees; damages sign boards			
F-1	32–50	70–110	Moderate tornado
The lower limit is the beginning of hurricane wind speed; peels surface off roofs; mobile homes pushed off foundations or overturned; moving autos pushed off roads; attached garages may be destroyed			
F-2	51–70	111–155	Significant tornado
Considerable damage. Roofs torn off frame houses; mobile homes demolished; boxcars pushed over; large trees snapped or uprooted; light object missiles generated			
F-3	71–92	156–204	Severe
Roof and some walls torn off well constructed houses; trains overturned; most trees in forest uprooted			
F-4	93–116	205–256	Devastating
Well-constructed houses levelled; structures with weak foundations blown off some distance; cars thrown and large missiles generated			
F-5	over 117	over 256	Incredible
Strong frame houses lifted off foundations and carried considerable distances to disintegrate; automobile sized missiles fly through the air in excess of 100 metres; trees debarked; steel reinforced concrete structures badly damaged			

There is in fact at least one more level of tornado intensity but it usually remains unquoted because it cannot do any more damage than the F-5.

immortalised in *The Wizard of Oz,* they are best-known for being an aspect of violent weather across parts of the US.

Spotting tornadoes

There is a modern network of Doppler radars covering the 48 contiguous states of the US. One of their purposes is to detect the rotating air marking the possible site of a developing tornado (which may occur at middle levels, a few kilometres above the surface) and to track it once it has reached the surface, again by mapping the signs of rotation. Doppler radar maps precipitation and other small particles and the size of the component of wind on which they are drifting – towards or away from it. As precipitation moves towards or away from the antenna, the received radar pulse will change in

Figure 5.4 Forecaster at the Storm Prediction Center, Norman, Oklahoma, discussing a radar image of a severe storm (Courtesy of NOAA/NWS).

frequency (Figure 5.4). If they are caught up in a tornadic circulation, it means that from the viewpoint of a Doppler radar, some will be moving towards its location and some moving away from it. This makes it possible for this type of radar to determine – by these differences in the flow components – what the detailed flow pattern of the wind is. Because the radar can scan at different angles above the horizon (it has a circular scan pattern for this purpose), it can sense the presence of mid-level rotation before anything happens on the ground. This higher-up spin does not guarantee the touchdown of a 'twister', but it is a very useful danger warning. Doppler radar helps the forecaster to concentrate on those severe storms most likely to become tornadic (Figures 5.4 and 5.5).

Members of the public provide a very valuable service. They volunteer observations to the *Skywarn* organisation, a network of storm spotters who work with their local community in order to take proper action if a tornado threatens. Skywarn observations are relayed immediately to the National Weather Service (NWS) where severe weather is monitored by Doppler radars, automatic weather station networks, weather satellites and other

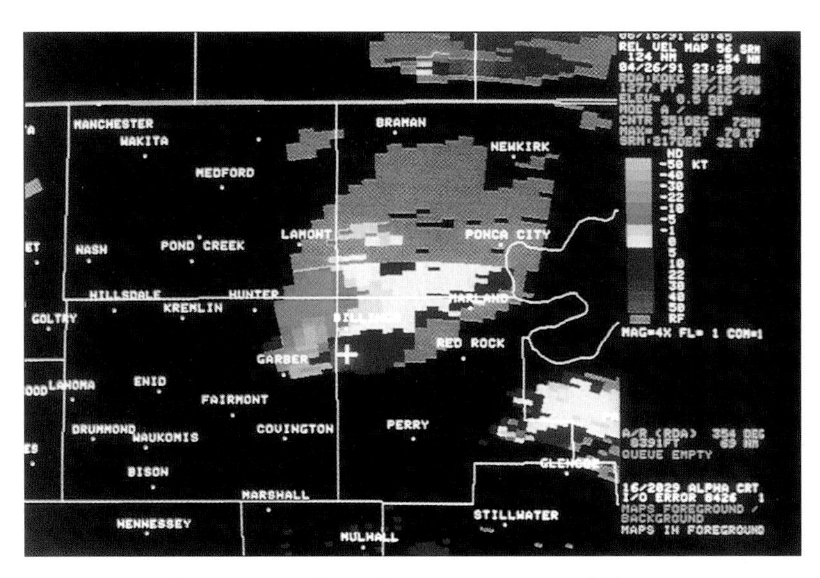

Figure 5.5 Doppler radar map of tornado storm activity over Oklahoma.

means. The NWS issues timely warnings of tornadoes to the public, usually via television and radio.

There are, of course, a whole host of amateur and professional spotters active in the field when tornadoes are about. In Oklahoma City, for example, the three local television stations all have their own response teams who drive out to look for the twisters. They vie with one another to get the first shot of touchdown and reports of the nature and extent of damage in the metropolitan area and beyond. One station has its own Doppler radar, issuing frequent updates, virtually town-by-town, of the most likely time of passage of tornado storms. The author has watched broadcast shots from another station's traffic helicopter, looking for signs of tornadic cloud from a reasonably safe distance! Television stations in the area send out chase cars to transmit live video footage of severe weather phenomena either locally or further afield in the state of Oklahoma and the same goes for TV stations in Texas, Kansas. Missouri and other states at high risk.

Members of the public are also on the lookout for drama and presumably excitement. However, chasing tornadoes, or severe storms in general, is a potentially dangerous pursuit. People who embark on such adventures should be absolutely certain what to do if they manage to get very close to or into such weather.

Scientists are working to understand much more about the origin and development of tornadoes and, indeed, there is much yet to learn. A major concentration of US expertise is based in Norman, 20 miles south of Oklahoma City. The University of Oklahoma houses a large School of Meteorology, and research into tornadic and other severe convective storms are one of its areas of excellence. Norman is also the home both of the National Severe Storms Laboratory (NSSL) and the Storm Prediction Center; the former concentrates on research, the latter on the nitty-gritty of day-to-day forecasting.

Both the University and NSSL run scientific field programmes to measure tornadic events. In the mid 1980s, NSSL used a 180 kilogramme (400 pound) drum packed with various weather sensors in the hope that, if placed judiciously in the track of a tornado, it would be able to obtain a series of direct observations. The package was called *TOTO* for 'TOtable Tornado Observatory' – more than coincidentally given the same name as Dorothy's dog in the *Wizard of Oz*. They did not succeed in a perfect crossing by a tornado but did gather data in the close vicinity. The TOTO drum now rests in the National Oceanic and Atmospheric Administration's museum in Silver Springs, Maryland.

By the mid-1990s, a new, intensive observational programme was created, it was *VORTEX*, an acronym for 'Verification of the Origins of Rotation in Tornadoes EXperiment'. The aim is to collect data in and around tornadoes with chase teams involving:

- a field co-ordinator
- photography
- mobile sounding with balloons
- mobile mesonets
- 'turtles'
- mobile and airborne Doppler radars.

A *mesonet* is a network of automatic surface weather stations deployed to provide frequently-sampled observations from small areas. A *turtle* is a flat, small, very solid sensor that can be stuck into the ground ahead of an approaching tornado.

How do tornadoes form?

Atmospheric scientists are still a long way from understanding fully why some supercell storms become tornadic while others which are apparently similar do not. Perhaps the differences are very subtle and it will take a lot more painstaking research to unravel them. For quite some years the large-scale atmospheric conditions favouring the development of supercell storms have been well known and, indeed, generally well predicted by the NWS. A very much more thorny problem is how to predict – even a few hours ahead – the development, intensity, track and longevity of individual tornadoes.

It is suspected that one critical factor is the existence at low-levels of boundaries that are the leading edges of cool air pools remaining from past thunderstorms. These small-scale 'fronts' may have a horizontal temperature contrast which sets up an invisible rotation about an elongated horizontal axis, rather like a rolling pin (Figure 5.6). The same horizontal rolling can be related to wind shear in the layer just above the surface, within which the wind speed increases with height.

Another important factor is an approaching thunderstorm that has an updraft supplying it. When the updraft impinges on the horizontal 'rolling' tube of air, they can interact in such a way that gradually tilts the roll upwards so that the vortex ends up vertically aligned like a top and is often intensified by stretching. All this may impart rotation to the whole storm updraft, a necessary but not wholly sufficient condition for a tornado to form. Rotation within the storm cloud can occur over areas typically 3–10 km (2–6 miles) across and is related to a small-scale low pressure minimum below and within the storm known as the *mesocyclone* into which air swirls with an anticlockwise motion (Figure 5.7).

Yet another apparently critical ingredient is the nature of the 'rear-flank downdraft', a feature that subsides at the trailing edges of supercell storms. This descending air is often heavily rain-laden and draws rotation down from aloft as it moves towards the surface (Figure 5.8). It wraps its way into the main storm updraft in the shape of a hook – this is sensed by precipitation radars as a characteristic feature in the precipitation pattern of tornadoes.

Once the descending air is trapped in the strong cyclonic circulation, rotation is intensified and focused towards a confined centre that can then develop into a funnel cloud. It has only fairly recently been appreciated that this 'concentration' occurs within a very confined area, probably just a few kilometres across. Teams from the University of Oklahoma and NSSL have therefore attempted to concentrate their field efforts on this region near to and inside the hook; they call it the *Bear's Cage*!

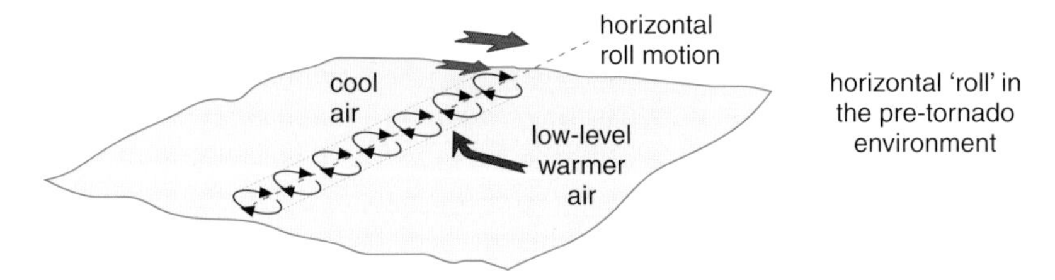

Figure 5.6 Cold outflow rolling horizontally.

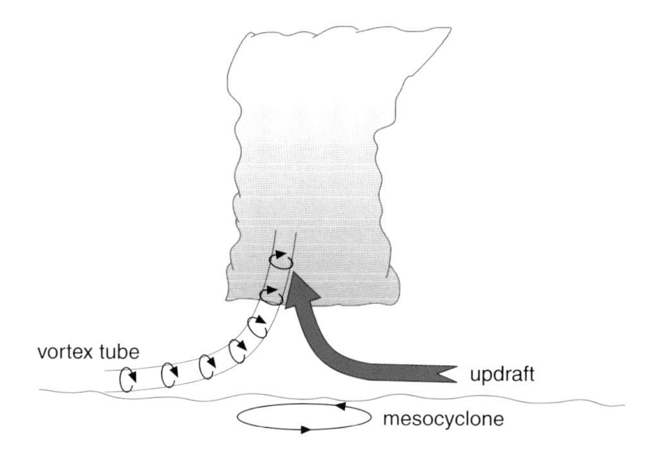

Figure 5.7 Tilted updraft and mesocyclone.

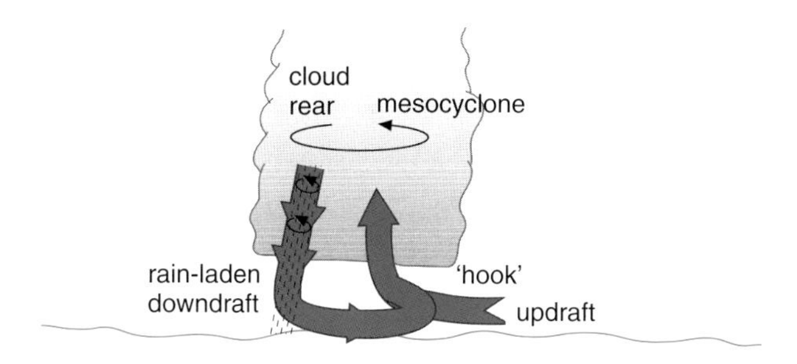

Figure 5.8 Rear flank downdraft in a severe convective cloud.

One major aim of this work is to decide if the nature of the rear-flank downdraft determines whether a tornado will or will not develop. If relatively 'warm', the descending downdraft air tends to be fairly buoyant and may rise when caught up in the concentrated rotation. If 'cold', the downdraft may however spread out and away from the storm. If downdrafts do become caught up in the rotation, they may dampen down the tornado which uses up a lot of its energy lifting the dense air. One thing is sure: the presence of a mesocyclone in itself does not guarantee the appearance of a tornado.

US tornado statistics

The peak season for tornadoes in the US is usually from March to June in the southern High Plains states and slightly later further north: from the Dakotas through the Great Lakes states to New England, the average season peaks either between May and July or between June and August. In the years from 1972 to 1991, on an annual average basis, tornadoes were ranked third as a cause of severe weather fatalities – behind floods and lightning but ahead of hurricanes. The average annual number of deaths from tornadoes was 69, a number varying a lot from year to year and displaying a gradual decrease over time (Table 5.2).

An important factor in the declining number is the increased understanding of tornado genesis and development, better monitoring and prediction, better warnings, better communication to the public and enhanced public awareness of the dangers involved, coupled with what to do when it happens. The location of people when they were killed by tornadoes gives some indication of relative risk (Table 5.3 shows data for the total number

Table 5.2 US fatalities caused by tornadoes

Period	**US fatalities annual average**
1940–49	154
1950–59	135
1960–69	94
1970–79	99
1980–89	52
1990–91	46

Table 5.3 Where tornadoes kill people

Location	**Facilities Number**	**As % of total**
mobile homes	230	38
'permanent' homes	162	27
vehicle	66	11
business	27	4
outdoors	56	9
school	15	2
other	49	8

and mean percentage of deaths in the USA caused by tornadoes for the years 1985–97 inclusive).

Not surprisingly, mobile homes are extremely susceptible to severe damage by tornadic storms, though the most violent can wreak havoc to much more substantially built structures.

The average annual total of US tornadoes is 761 for the period 1950–94. The average monthly totals emphasise their strong seasonality (Table 5.4). These data are a reminder that although there is an obvious summer peak, tornadoes do occasionally occur in the depth of winter.

Where are tornadoes most likely?

The average number of tornadoes per year across the US between 1989 and 1998, categorised by state, reveals a broad picture of their geographical distribution. The number for each state needs to be normalised by dividing it by the area of the state to get some feel for the 'density' of tornado occurrence; calculations on that basis give the area which witnessed one tornado per year for the top ten states (Table 5.5). However, these figures mask important detail. It may be surprising to see Maryland higher up the scale than Texas, or West Virginia above Oklahoma. One factor is the particular period of this data – 10 years during which there were some important outbreaks in the northern Appalachian and Chesapeake regions. Looking at the same statistics, but for the 1950 to 1998 period, Oklahoma 'persists' with one tornado in an area of 3,422 square km, Texas has one in 5,542 square km, while Maryland slips to a value of one in 6,648 square kilometres and West Virginia plummets to one every 31,400 square km.

The deadliest tornadoes of all

The ten most deadly tornadoes to hit the USA all occurred at least 50 years ago (Table 5.6), a tribute to all the scientific and technical improvements that have been achieved over recent decades.

An active year

The year 1998 saw an anomalously high tornado death toll of 130, concentrated mainly across Florida, Alabama and Georgia. Two tornadoes were

Table 5.4 Mean monthly frequency of US tornadoes

Jan	Feb	Mar	Apr	May	Jun	Jul	Aug	Sep	Oct	Nov	Dec
13	21	51	102	163	160	88	58	37	23	28	17

Table 5.5 The top ten US states for tornado frequency

Rank	State	Area per tornado 1989–98 (square km)
1.	Florida	1,890
2.	West Virginia	2,088
3.	Indiana	2,152
4.	Maryland	2,452
5.	Kansas	2,797
6.	Oklahoma	3,242
7.	Iowa	3,262
8.	Louisiana	3,344
9.	Illinois	3,345
10.	Texas	4,058

Table 5.6 The ten worst tornadoes in the US

Rank	Date	Location	Deaths
1.	18 March 1925	Missouri, Illinois, Indiana	689
2.	6 May 1840	Natchez, Mississippi	317
3.	27 May 1896	St Louis, Missouri	255
4.	5 April 1936	Tupelo, Mississippi	216
5.	9 April 1936	Gainesville, Georgia	203
6.	9 April 1947	Woodward, Oklahoma	181
7.	24 April 1908	Amite, Louisiana and Purvis, Mississippi	143
8.	12 June 1899	New Richmond, Wisconsin	117
9.	8 June 1953	Flint, Michigan	115
10.	11 May 1953	Waco, Texas	114
11.	18 May 1902	Goliad, Texas	114

assessed to have reached the rare F-5 intensity – on 8 April in Pleasant Grove, Alabama and on 16 April at Waynesboro, Tennessee.

During the night of 22–23 February a crop of tornadoes swept across central Florida – an unusual event during the hours of darkness. On 22 February a convective disturbance moved eastwards across the Florida panhandle arriving in the north-eastern part of the state during the morning and afternoon. Low-level outflows were generated, one of which slowed down across north central Florida. At the same time, a marked squall moved into the area from the Gulf of Mexico. South-south-easterly winds at low levels were capped by westerly winds that strengthened in the middle and upper troposphere.

All this meant that the larger-scale synoptic flow displayed the all-important shear within the troposphere and, coincidentally, the development of a lower-level environment favouring convergence boosted by strong surface heating. All these ingredients added up to the genesis of tornado-producing deep convection.

The onset of the severe weather occurred at about 23:00 local time round the cities of Kissimmee and Orlando – places well known to millions of holidaymakers to Florida. Thunder, lightning and intense rain were the herald of disaster: two hours later, 36 people had been killed by tornadoes. It was the worst outbreak in the history of the state, inflicting more fatalities than Hurricane Andrew. The night-time menace meant that many residents were not as well prepared as if it had been a more typical daylight event. Hundreds of homes were destroyed and, after a few weeks, the death toll had reached 42.

The NWS Forecast Office in Melbourne, just about in the middle of the east coast of Florida, was responsible for issuing warnings of severe weather conditions. Warnings for that night are instructive. In the middle of the evening came the following:

> *TORNADO WARNING*
> *NATIONAL WEATHER SERVICE MELBOURNE FL*
> *945 PM EST SUN FEB 22 1998*
>
> *THE NATIONAL WEATHER SERVICE IN MELBOURNE FL HAS ISSUED A TORNADO WARNING EFFECTIVE UNTIL 1050 PM EST FOR PEOPLE IN THE FOLLOWING LOCATION IN EAST CENTRAL FLORIDA . . . LAKE COUNTY.*
>
> *AT 945 PM . . . DOPPLER RADAR INDICATED A POSSIBLE TORNADO APPROACHING WILDWOOD IN SUMTER COUNTY. THE STORM CONTAINING THE POSSIBLE TORNADO WILL MOVE INTO WESTERN LAKE COUNTY NEAR LAKE GRIFFIN AND WILL MOVE ACROSS LADY LAKE AND EMERALD BETWEEN 955 AND 1015 PM.*
>
> *IF YOU ARE IN THE PATH OF A TORNADO . . . ABANDON CARS AND MOBILE HOMES FOR A REINFORCED BUILDING OR GET*

INTO A DITCH OR CULVERT. THE SAFEST PLACE IS AN INTERIOR ROOM SUCH AS A CLOSET ON THE LOWEST FLOOR OF A STRONG BUILDING. AVOID WINDOWS.

and in the middle of the night:

TORNADO WARNING
NATIONAL WEATHER SERVICE MELBOURNE FL
304 AM EST MON FEB 23 1998

THE NATIONAL WEATHER SERVICE IN MELBOURNE FL HAS ISSUED A TORNADO WARNING EFFECTIVE UNTIL 400 AM EST FOR PEOPLE IN THE FOLLOWING LOCATION IN EAST CENTRAL FLORIDA . . . OSCEOLA COUNTY.

AT 300 AM . . . DOPPLER RADAR SHOWED A SEVERE THUNDERSTORM AND TORNADIC CIRCULATION DEVELOPING JUST EAST OF THE SOUTH END OF LAKE KISSIMMEE. THE STORM AND POSSIBLE TORNADO WILL MOVE NORTHEAST AT 45 MPH THROUGH LAKE MARIAN . . . AND ACROSS THE FLORIDA TURNPIKE TO NEAR KENNANSVILLE BETWEEN 305 AND 320 AM.

IF YOU LIVE NEAR THE STORM . . . TAKE SHELTER IMMEDIATELY!!! TORNADOES HAVE ALREADY CLAIMED A HALF DOZEN LIVES IN CENTRAL FLORIDA THUS FAR.

IF YOU ARE IN THE PATH OF A TORNADO . . . ABANDON CARS AND MOBILE HOMES FOR A REINFORCED BUILDING OR GET INTO A DITCH OR CULVERT. THE SAFEST PLACE IS AN INTERIOR ROOM SUCH AS A CLOSET ON THE LOWEST FLOOR OF A STRONG BUILDING. AVOID WINDOWS.

American F-5 tornadoes

Fortunately category F-5 tornadoes are rare. Their number varies significantly from year to year; Table 5.7 summarises the NWS's best estimate of their incidence over the US during a recent period. Absent years means that no F-5 systems were observed.

The Oklahoma outbreak of May 1999

The largest outbreak in Oklahoma history hit the state during the afternoon and evening of 3 May 1999. More than fifty tornadoes were reported during that day in central Oklahoma (Figure 5.9) during which 40 people lost their lives. There were in fact even more tornadoes across a wider region which included south-central Kansas, eastern Oklahoma and northern Texas. Five

Table 5.7 Incidence of F-5 tornadoes in the US

Year	No. of tornadoes
1973	1
1974	7
1976	3
1977	1
1982	1
1984	1
1985	1
1990	3
1991	1
1992	1
1996	1
1997	1
1998	2
1999	1

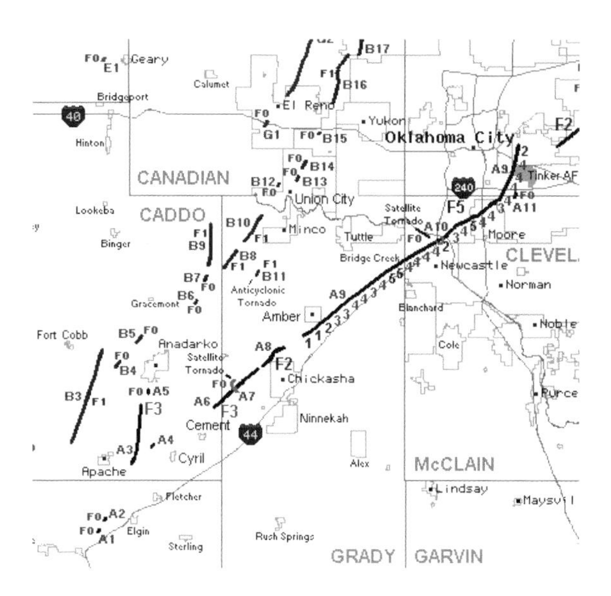

Figure 5.9 Track and F-rating of tornadoes on 3 May 1999 in and approaching Oklahoma City. All tornadoes travelled north-eastwards.

people died, over 100 were injured and heavy damage occurred in Wichita, Kansas.

Television channel tornado chasers and those from the University of Oklahoma and NSSL in Norman were out for reasons of public safety – relaying information to the NWS and TV broadcasters, or for research – taking advantage of the possibly once-in-a-lifetime event to occur right on their doorstep. It quickly became obvious to everyone involved that the intensity of the tornadoes they were monitoring was something very unusual. As the severe weather closed in on the south-west outskirts of Oklahoma City, one particularly huge 1.5 km (1 mile) wide tornado rampaged through several suburban communities including that of Moore. It was an F-5!

The last such tornado to hit Oklahoma was at Broken Bow in the south-eastern part of the state on 2 April 1982. The state capital itself had never experienced such an event: the previous most deadly to have hit happened on 12 June 1942, again in the south-western part of the city. Thirty-five people were killed and $500,000 worth of damage occurred to 70 homes.

The 1999 May outbreak was far more devastating. In Oklahoma and Cleveland counties (essentially Oklahoma City and the areas to its south), 1,780 homes were utterly destroyed, 6,550 homes were damaged and 85 businesses were flattened as were 3 churches and 2 schools. In neighbouring counties a further 534 homes, 79 businesses and 2 churches were demolished by the winds.

The damage was incredible (Figure 5.10) – brick houses were razed to the ground over large areas; the scene resembled the total destruction caused by an atomic bomb blast. Even though timely warnings were issued, the winds were so strong that some people who sheltered as properly as they could within their homes perished nevertheless. Many of the newer houses in this area did not have basements in which to shelter; even if buried by the house collapsing on them, basements are safer than ground-level rooms.

Trouble for central Oklahoma started that afternoon in Comanche and Caddo Counties to the south-west of Oklahoma City. Two supercell storms spawned 11 relatively short-lived tornadoes with the strongest achieving F-3 between Apache and Anadarko. There was no damage in Comanche County; in Caddo County one home was destroyed, one was badly damaged and one received minor damage in the small community of Stecker. As the afternoon wore on, the supercell complex that had produced this F-3 gave birth to the real trouble just inside south-east Caddo County. Another tornado, running north-eastwards into neighbouring Grady County, attained F-3 status in the vicinity of Chickasha. Here, two hangars and four other buildings at the airport were destroyed. In the county's rural areas, 75 homes were destroyed and 25 were damaged.

Figure 5.10 Extreme damage in Moore, Oklahoma after 3 May 1999 tornado (Courtesy of Dr R Peppler, University of Oklahoma).

After this F-3 abated, the tornado that was to produce the worst damage in the history of the state touched down as an F-1 just north-east of Chickasha. It, too, ran north-east, directly towards the metropolitan area of Oklahoma City; the alarm bells already ringing increased as the winds within the funnel cloud wreaked more and more havoc. It increased in intensity from F-1 as it tracked parallel to the H.E. Bailey Turnpike (Interstate Highway 44) which runs north-east into Oklahoma City from Chickasha. The evidence suggests that the tornado strengthened to an F-4, dropped back to an F-3, then re-intensified to an extremely rare F-5 as it crossed the community of Bridge Creek. It was here that 680 homes were destroyed.

Next the tornado moderated to an estimated F-4 as it crossed into McClain County, north-east of Newcastle. In this area, 30 homes and mobile homes were destroyed, 80 sustained major damage and 40 more minor damage. It seems then that, while this one declined, another was spawned very close to the Canadian River and crossed into Cleveland County. It was now evening as the winds rapidly attained F-5 status with the tornado rampaging north-eastwards. Unfortunately for the citizens of Moore it was their community where this happened. There, 1,225 houses and 274 apartments were destroyed and between 4,000 and 4,500 were damaged. Fifty businesses were flattened as well as two churches and two schools.

It weakened to a still dangerous F-3 in north-east Moore before running into Oklahoma County, across the Tinker Air Force Base and the neighbouring communities of Del City and Midwest City (eastern suburbs of Oklahoma City) late at night. Del City and Midwest City suffered 569 and 188 homes destroyed respectively while some hotels and churches were assessed as structurally unsafe. In Oklahoma City, 633 homes and 152 apartments were destroyed. A church and 18 businesses were demolished.

And so it went on – other tornadoes ran across counties to the west and north, and to the east of Oklahoma City. Estimates run to a total of nearly a billion dollars of damage, outstripping the previous most costly ($884 million) which occurred in 1979 in Wichita Falls, north Texas.

Aid for people affected

Some 9,500 affected residents applied for federal and state aid, the majority from Oklahoma, Cleveland and Grady Counties; more than $67.8 million worth of grants and low-interest loans were authorised. Of that total, over $61.7 million (approved by the US Small Business Administration) was in the form of low interest loans to homeowners, renters and owners of businesses. In excess of $1.7 million came from the Federal Emergency Management Agency for smaller emergency home repairs and short-term rental assistance while $121,000 was paid to people unable to work because of the disaster.

After the tornadoes, 1,600 people sought refuge in 10 disaster shelters in the Oklahoma City area. Most found other accommodation quickly, with relatives, or temporarily in hotels or in apartments. Most owner-occupiers had quickly received financial support from their property insurers but it would take a year or so before the new houses were constructed. For the uninsured, the American Red Cross paid hotel bills plus the first month's rent for an apartment.

That outbreak would most certainly have led to many more fatalities, perhaps 700 or more, had it struck in 1949 not 1999; no tornado since 1953 has led to more than 100 deaths. Present observational and forecasting methods contrast starkly with those employed by officers of the Army Signal Corps in the early 1800s when it was estimated that there might be about 50 tornadoes annually in the central part of the USA. This, of course, was a gross underestimate but still is believed to have scared some settlers into not joining the great sweep west to settle in the threatened areas. During this period the word 'tornado' was apparently banned from forecasts!

Another critical time was in March 1948 when, on the 25th, two members of staff at Tinker Air Force Base in Oklahoma City issued the first successful tornado forecast. These were seemingly only for military ears, so the US Weather Bureau (now the National Weather Service) began to predict severe storms in 1952. All the scientific and technical changes since then bring us up to date – to today's excellent service.

The damage

Tornadoes can destroy really substantial buildings; the damage they cause is due both to the dynamic force of their massively strong winds and to the flying debris. Debris often consists of sticks and branches, glass, various building materials and, in extreme cases, cars, farm animals and larger household appliances. In a 1975 tornado in Mississippi, a home freezer was carried for a mile before landing. Virtually all of this debris is deadly, flying through the air at high speed.

One of the ways buildings are destroyed is by the wind and its embedded debris blowing down a wall. This lifts the roof, after which the other walls fall outwards. Such damage is not caused by the structure exploding – the notion that windows should be opened to lessen this risk is nowadays quite properly discounted.

Warnings

As far ahead as feasible, warnings are issued usually to cover the area of several counties within a state. The public will normally be aware of the general risk of tornadic thunderstorms either the previous evening or early in the morning when they hear the breakfast news and weather forecasts. As the day progresses, the main risk areas become clearer as spotter reports and other information start to come in.

The basis of the warning is the location and size of the 'parent' storm together with its direction and speed of movement, both of which can be erratic.

Safety rules

What should you do if you are in a threatened area? The very best advice is to leave but not everyone can do that! The NWS guidance rules on safety during tornado outbreaks are:

- *At home*: the basement offers the greatest safety. Seek shelter under furniture if possible. In homes without basements, take cover in the centre part of the house, on the lowest floor, in a small room such as a closet or bathroom, or under sturdy furniture. Keep away from windows.
- *In shopping centres*: go to a designated shelter area (not to your parked car).
- *In office buildings*: go to an interior hallway on the lowest floor or to a designated shelter area.

- *In schools*: follow advance plans to a designated shelter area, usually an interior hallway on the lowest floor. If the building is not of reinforced construction, go to a building that is, or take cover outside on lower ground. Stay out of auditoriums, gymnasiums, and other structures with wide, free-span roofs.
- *In cars*: leave your car and seek shelter in a substantial nearby building, or lie flat in a nearby ditch or ravine.
- *In open country*: lie flat in the nearest ditch or ravine.
- *Mobile homes*: are particularly vulnerable and should be evacuated. Trailer parks should have a community storm shelter and a warden to monitor broadcasts throughout the severe storm emergency. If there is no shelter nearby, leave the trailer and take cover on low, protected ground.

The risk in Britain

Tornadoes are more common in the UK than one might expect. Although nowhere near as severe as the American outbreaks, they are not uncommon. They are the real thing, with funnel clouds and some damage at the surface, and with an occasional indication of something resembling a mesocyclone.

Like their US counterparts, they are always related to very deep convection accompanied by hail and very heavy rain. On average, they are observed on about 20 days a year over the UK, peaking in June, July and August. The British Isles does not however have the geographical features which lead to such terribly severe systems in the most susceptible region of the US. Just as in the US, 70% or so of reported tornadoes in Britain occur in the afternoon.

An alternative intensity scale

The UK Tornado Research Organisation (TORRO) devised its own intensity scale to categorise not only events affecting the British Isles but also elsewhere. It differs from the Fujita scale used in the US. Table 5.8 lists the first five categories of tornado likely to affect the British Isles.

Perhaps the most remarkable recorded UK outbreak was on 23 November 1981 when 105 tornadoes were observed. Two more recent events occurred over southern England; coastal tornadoes were reported near Selsey on the Sussex coast late at night on 7 January 1998, while the outskirts of Reading were struck in the late afternoon of 13 June of the same year.

In the January episode, a series of relatively small-scale but very active troughs moved west to east along the Channel coast. They were coherent features composed of thunderstorms producing localised and extremely intense rainfall (Figure 5.11). The particular feature spawning the Selsey tornado could be tracked back over the previous few hours to Dorset and

Table 5.8 UK Tornado Research Organisation intensity scale

Intensity	Wind speed metres/second	mph	Type
T0	17–24	39–54	Light tornado
Loose light litter raised from ground level in spirals. Tents, marquees seriously disturbed; most exposed tiles, slates on roofs dislodged. Twigs snapped; trail visible through crops.			
T1	25–32	55–72	Mild tornado
Deck chairs, small plants, heavy litter made airborne; minor damage to sheds. More serious dislodging of tiles, slates, chimney pots. Wooden fences flattened. Slight damage to hedges and trees.			
T2	33–41	73–92	Moderate tornado
Heavy mobile homes displaced, light caravans blown over, garden sheds destroyed, garage roofs torn away, much damage to tiled roofs and chimney stacks. General damage to trees, some big branches twisted or snapped off, small trees uprooted.			
T3	42–51	93–114	Strong tornado
Mobile homes overturned/badly damaged; light caravans destroyed; garages, outbuildings destroyed; house roof timbers considerably exposed. Some of the bigger trees snapped or uprooted.			
T4	52–61	115–136	Severe tornado
Mobile homes destroyed; some sheds airborne for considerable distances; entire roofs removed from some houses or prefabricated buildings; roof timbers of stronger brick or stone houses completely exposed; possible collapse of gable ends. Numerous trees uprooted or snapped.			

T0 to T3 are designated 'weak' tornadoes while T4 is the lowest grade of those defined as 'strong'.

further west. One ingredient in the tornado's formation may have been the much warmer-than-average sea surface of the Channel, 2–3°C higher than normal. Local reports of hail that increased from marble to golfball size over a matter of a couple of minutes attested to the severity of the convection – the hail was followed a minute or so later by the tornado itself which was probably F-1 or F-2 on the Fujita scale; subsequent analysis of the damage suggested T3–T4 on the TORRO scale.

The likely influence of warmer-than-average sea water in the English Channel during autumn 1997 and the following winter is supported both by the Selsey event and by a tornado which smashed through the small communities of Atherfield and Kingston on the Isle of Wight around midday on 4 January 1998; it started life as a water spout over the sea. Preliminary

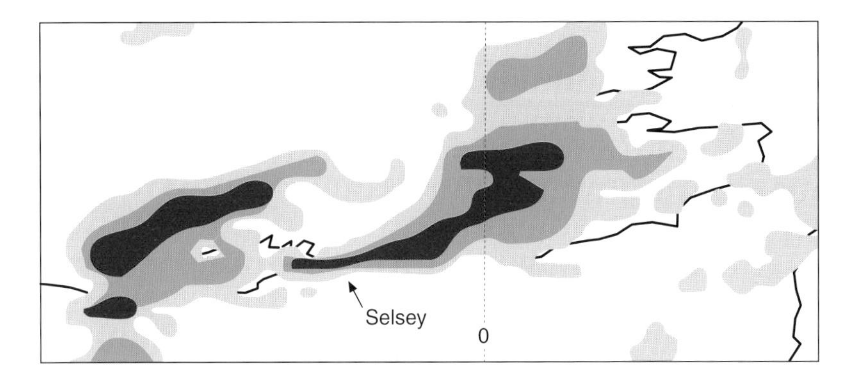

Figure 5.11 Radar-mapped rainfall at 00:00 UTC, 8 January 1998.

analysis suggests that there were perhaps 55 UK tornadoes in 1998; all were in the 'weak' category on the TORRO scale but even the small-scale tornadoes in Britain can cause great distress for those in their paths.

Index